Stone Cladding Engineering

Rui de Sousa Camposinhos

# Stone Cladding Engineering

Springer

Rui de Sousa Camposinhos
School of Engineering
Department of Civil Engineering
Instituto Politécnico do Porto
Porto, Portugal

ISBN 978-94-007-6847-5 ISBN 978-94-007-6848-2 (eBook)
DOI 10.1007/978-94-007-6848-2
Springer Dordrecht Heidelberg New York London

Library of Congress Control Number: 2013943407



Printed on acid-free paper

Springer is part of Springer Science+Business Media (www.springer.com)

# Preface

The natural stone is a material with widespread use over time, mainly due to its robustness, durability and availability in a variety of colors and textures. If, for many centuries, it was practically the only structural material for enduring longstanding construction, today, with the appearance of new construction materials, especially reinforced concrete, its use has been progressively dedicated to dressing of building façades. Under this usage, technological developments have enabled the delivery of elements of natural stone cladding with ever larger dimensions and smaller thicknesses, contributing to economic competitiveness in its use, particularly when assessed in the complete lifecycle of construction. The progressive increase in stone plate size demands new requirements regarding the capacity to withstand forces, notably wind and seismic ones and the attachment system, crucial in view of the potential damage that a detachment of a high weight plate can cause. These elements must therefore be the subject of a structural design to ensure an appropriate level of safety. As the dimensioning of this type of elements is not currently explicitly regulated, existing regulation principles should be adapted, especially those covered in the recent European Structural Regulations, the EuroCodes or also under the Uniform Building Code in the USA. It is in this context that this book, *Stone Cladding Engineering*, finds its place, providing a vast pool of scientific and technical knowledge that, in conjunction with the current regulatory framework, allows the design of cladding in natural stone.

The author of the book, Professor Rui Camposinhos, has devoted to this subject particular attention in recent years promoting experimental studies involving the execution of hundreds of pullout tests in order to characterize the behavior of various types of natural stone and guiding research, which resulted in the publication of papers on the subject in prestigious international journals. The extent to which the subject is treated also results from his academic studies, obtaining a master's degree and then a Ph.D. in related topics, and his experience as a professor at the School of Engineering of Polytechnic Institute of Porto and research work developed for several years, integrated into research centers of the Faculty of Engineering of the University of Porto, such as the project "Presstone", to develop a prototype "Natural Pre-stressed Stone Façade Systems" of which he is a coordinator, should also stand out.

The material of the book is well organized, allowing the reader to follow the various threads in the proper sequence, well written and clearly illustrated, facilitating the understanding of the various aspects associated with the theme, the dimensioning and design of natural stone cladding plates, which can be considered divided into three parts. The first part organized in three chapters, after providing an introduction to the subject in a separate chapter, presents the characterization of natural stone, sorting them by type and identifying their main physical and mechanical properties. A detailed chapter is dedicated to the discussion of several wall and cladding systems.

The second part, organized in two chapters, presents the key concepts and methodologies for verification of structural safety, checking the limit conditions involved, the main forces considered and the concepts of reliability and safety factors involved in a form of safety procedures according to modern structural regulations. In this part a chapter is also dedicated to the characterization of actions in façades and its characteristic and design values. Special emphasis is given to determine the effects of seismic action, which shows in detail a general methodology applied to non-structural elements linked to structures. The chapter ends with methodologies for the determination of flexural and tensile stone strength.

In the third part, three chapters present the matters that are necessary for proper sizing and detail of particular solutions in relation to determining the resistance of the plates, the calculation of kerf, undercut and dowel and pin anchorages systems. Each chapter ends with examples of applications, illustrating the main topics exposed and presenting a comprehensive analysis of the solutions. The last chapter is opportunely devoted to the design of stainless steel and aluminum alloys body anchors and rails.

The way the content of this work was developed might be the basis for a discipline to be integrated to a Master's degree that would be of great benefit to both students and engineering technicians, who wish to obtain or deepen their knowledge on this topic, and will certainly constitute an indispensable tool for all engineering professionals involved in the design or application of natural stone slabs for façades.

Porto, Portugal

Raimundo Delgado
(Full Professor FEUP)

# Contents

# Chapter 1
# Cladding with Stone: Introduction

**Abstract** External cladding of buildings in modern architecture has the important function of protection against weathering. Aesthetic requirements with needs related to the durability of the different elements require that building's veneers act like skins undergoing mechanical, chemical, thermal, and hygrometric stress related to human activity, the quality of the air, and weathering.

Following modern structural design rules and technologies installation procedures and design assumptions are presented, mainly, to the dowel anchorage, the undercut anchorage technology and to the kerf anchorage system.

Attention is called to the state of art as far as façades stone cladding is concerned which is not updated according to the recent evolution verified for the "man-made" materials design approaches. This way attention is paid to the fact that the time has come to move forward in step with "new" design and calculation methods that efficiently and safely meet the needs that must be overcome by the natural stone sector.

## 1.1 Introduction

Cladding with natural stone certainly has its origin in the way how our civilization started to build its shelters, i.e., stone by stone. The stonework developed in a skilful way throughout the times with some well kept "secrets" transferred trough generations. The facing appearance of the façades is very similar to that of the former stoned walls, though the stereotomy is quite the same.

The most ancient remaining stone structures are scattered throughout the world: from the Egyptian pyramids, Maya, Peruvian, Greek and Roman temples, to the castles, churches and monuments that flourished during the high and late medieval period from Romanesque and later by the Renaissance architecture (Fig. 1.1).

Through the last centuries, progress in technology has contributed to significant alteration in the use of stone. The industrial revolution changed this situation radically. During the nineteenth century, the worldwide tonnage of iron produced

R. de Sousa Camposinhos, *Stone Cladding Engineering*,
DOI 10.1007/978-94-007-6848-2_1, © Springer Science+Business Media Dordrecht 2014

**Fig. 1.1** A view of the Pisa Cathedral which first stone was laid in 1093, initiating what would become the distinctive Pisan Romanesque style

had increased by more than 50 times. Encompassing an impressive technological improvement all the impurities were burned out of the iron ore with precise amounts of carbon added for hardness leading to the origin of an impressive material: the steel.

This new material attained a remarkable tensile strength with capacities that would transfigure our architecture and structural design conception. Later steel appeared easily formed into long thin bars and enclosed in cheap, freely formed concrete resulting in a strong, economic, easily produced structural member which could take almost any form. Interesting to remark is that both steel and concrete have their origin in rock, though of different types, and are undoubtedly the most common building materials used today.

Again, a derived rock material, the glass, emerged with the Industrial Revolution. The glass industry had developed in the same direction permitting wide openings in the enclosure of the buildings.

Thus, exterior walls were dismissed from its main function as load-bearing element of the building. Their structural function was not necessary for the façade to be treated as a dressing that enfolded the structural components. This performance may be the justification for the word "façade" with its origin from the Italian *facciata*, or *faccia* meaning "face". This "skin" still needed to transfer wind loads to the frame, but it was no longer required to carry the floor loads.

## 1.2 Stone Veneers Performance

During the twentieth century buildings envelope used primarily brick or other smaller unit-type materials, but by the 1920s larger limestone slabs, typically at least 10 cm thick, began to be used with greater frequency. During this time and

**Fig. 1.2** A view of the Empire State Building west oriented façade

until the 1950s, each floor was typically designed individually with panels stacked vertically between supports near the floor levels and horizontal movement joints installed directly below the support or at mid-storey.

Empire State Building is a paradigm of its era; the structural steel frame incorporates two spandrel beams at each floor: an inner beam to receive the floor's live and dead loads and the outer beam to support the exterior wall system. The façade consists of a series of vertical bands of brick back-up masonry faced with limestone, alternating with vertical bands of steel-framed windows (Fig. 1.2).

The brick backup masonry is anchored to the structural steel columns with steel rods and the limestone is anchored to the brick masonry with flat section bent iron bars that are hooked into the brick masonry and into cut kerfs in the top, bottom and side edges of the limestone slabs [1].

Various methods are used to obtain the stone blocks from the quarries depending on the type, stratification and its scarcity or abundance. The main goal for the process is to obtain the largest possible flawless blocks, thus minimizing wastage. Nowadays, modern technology allows cutting the stone in precise dimensions with smaller and smaller thickness, thus providing effective cost solutions for fieldstone veneer.

Thickness of stone veneer on buildings has been significantly reduced during the last decades, this way, common thickness as thin as 20 mm may be found. Regrettably, this trend has not occurred within a scientific or technical supporting research, but due to economical interests.

One of the obvious consequences is the huge number of substantial failures that arise from using thin dimension stone cladding without base on evidence and sound research. That's why, most probably, research has been increasing during last years as testing and setting of standards are now encompassing the stone industry as it should be [2–5].

## 1.3 Cladding with Stone

External cladding of buildings in modern architecture has the important function of protection against weathering. Aesthetic requirements with needs related to the durability of the different elements, environmental compatibility and economic sustainability demand that building's veneers act like skins undergoing mechanical, chemical, thermal, and hygrometric stress related to human activity, the quality of the air, and weathering. These surfaces play an important role in both the aesthetic impact of the building and its integration into the surrounding landscape, and the materials must match the need for protecting the environment and for sustainable development.

Several methods or systems are available for installing stone on the exterior of buildings [6]. Their success will depend on the manner how the environment and exposure conditions are addressed. Temperature changes, air pressure, water, in any form, as direct or indirect induced tensions have to be properly undertaken, based on a solid empirical investigation encompassed with lessons from older thin-stone cladding building's façade systems.

Direct or adhered fixing systems consisting on cement bonding the entire back slab's surface has several drawbacks. Differential displacements and deformations under the same action between the slab's bed, the joint and the slab itself if prevented are the main causes for the well-known malfunctions and failures.

The fact that the backup and the cladding system are rigidly connected implies that the differential dimension changes are restrained giving rise to stresses in the cladding system: grout and slabs, with different distributions depending on the type and origin of the actions. This have to occur due to the fact that the physical and mechanical properties of the cladding and the grout are different, namely their coefficient of thermal expansion, hygroscopic coefficients and elastic modulus.

As a consequence stress concentrations are to be found in the periphery of the slabs and in the joints between them thus promoting adhesion failure of wall cladding with the inherent slab's detachment.

## 1.4 Design Methodology

One way to prevent water penetration is to avoid the direct fixing of the cladding to the wall leaving an air space between the cladding and the wall or substrate thus protecting the building from weathering and particularly from the infiltration of rainwater into the building walls and acting as a rain screen.

The mechanical anchoring of cladding besides being an effective fixing system gives way to an easy installation of rain screen systems either ventilated or pressure equalized [7], however cuts and holes are necessary to be made on the slab's thickness or backing. The most common cut shapes are the hole for dowell insertion, a kerf cut for angles or double T anchorages; slot cuts for disks with shanks and the cylindrical hole undercut for the insertion of a cone bolt with sleeve [8].

The structural detailing of any dimension stone cladding system involves the determination of the anchoring properties and the section of the inserts, the dimensions of the slab and the performance of the cladding system.

The design method based on allowable stresses has been and, it can be said, still is the most commonly used method for dimension stone. This method itself contains no rules for defining these allowable stresses.

Traditionally, safety factors have been applied by comparing the ratio of the "design" strength to the "design" stress (load) with standard or accepted values for the stone and/or the situation which is being designed. These safety values are subjective by taking into account the rupture stresses of the stone itself. There are very few recommendations about safety factors in dimension stone design and they all cover an allowable stress design approach [8–11].

Global safety factors recommended by some standards range are followed by stone industry associations, yet when comparing equivalent values in modern codes, although partial safety factors for limit state design are inferior to the so-called allowable stress design factors, this does not necessarily mean that the latter result in higher safety [12–14].

One reason stone safety factors are more conservative is that stone is a natural material and not a closely controlled manufactured product. The physical properties of stone, even for the same quarry, can vary widely. Some stones also lose strength after repeated heating-cooling and freeze-thaw cycles. Some gain or lose strength when they're saturated.

Traditional stonemasonry skills and knowledge must be retained and integrated into modern stonemasonry practice so that a depth and breadth of expertise is maintained. Stone buildings from past times provide examples of sought after skills that are rare, even non-existent today.

Most architectural specifications require that stone met certain specified, European and North American or other testing standards before it is accepted for use. For instances in the European market, the harmonized standards and CE marking have important consequences for producers. It is the responsibility of the supplier of the final natural stone product, to ensure that the properties of the product are documented before the product is sold or being used in a construction.

It must be emphasized that the major innovation in the European standards in contrast to the traditional national standards is that they are geared to evaluation of conformity and factory production control. It is now required that the supplier or manufacturer of dimension stone products shall be responsible for the assessment and attestation that the products offered are in conformity with the requirements of European standards. The compliance with the requirements of the published standards and stated values, initial type testing and permanent production control, with registration of result have to be made available.

In Canada and United States, it's generally required that dimension stone complies with specified ASTM or other testing standards before it is accepted for use.

Nevertheless, the stone has not yet followed in a comprehensive systematic manner the generalization of a reliability-based approach that is implicit in all countries for all construction materials.

Information and guidance has to be given to obtain the characteristic values of dimension stone strength based on the information that is mandatory to collect in some of the relevant standard tests.

In this context, based on the partial factor method according to the limit states design, structural calculation formulae for dimension stone slabs is presented together with installation procedures and specific detailing of the most well-known modern anchoring fixing systems.

## 1.5 Actions and Stone Strength

The value of the actions and their combination rules values are presented in order to determine their effects on building's façade and cladding systems.

Expressions are given to determine the self-weight of dimensioning stones considering the water absorption and open porosity of the corresponding natural stone slabs.

Formula and calculation examples are given to compute the external and internal pressures due to the wind action on pressure equalized, ventilated or joint sealed vented rainscreen claddings systems.

Simple formulae with application examples are presented to compute the maximum response of cladding elements attached to buildings façades when subjected to earthquake ground motions transmitted throughout the building structure.

Guidance is also given for the calculation of the joint's spacing and width regarding movements originated by thermo-hygrometric actions.

## 1.6 Fixing Systems and Metal Anchors

It has been shown that, for most situations, a design focused on the bending strength of stone panels, to the detriment of the anchorage capacity, is unsafe. Yet, normal practice, as mentioned above, does not take this into account.

In designing stone cladding systems with mechanical anchorage, two different effects must be taken into account when evaluating the effective stress in the critical region: the geometry and the stone specific properties.

The first factor (geometry) will cause stress concentration near the critical region that can be determined through analytical or numerical methods. The second factor (stone properties) further amplifies the stress, typical of quasi-brittle stones.

The stress magnification factor depends on the observed spall angle, and this angle is somehow related with the type of stone.

The use of the finite element method with linear elastic material properties, together with the maximum principal stresses failure criteria when module of rupture is attained, has been shown to be an appropriate design procedure for estimating the breaking load at the anchorages.

In this book installation procedures and design assumptions are presented to the dowel anchorage system where holes location on slabs edges is discussed taking into consideration the induced stresses by flexure under lateral actions [15].

Design assumptions are also presented to the undercut anchorage technology, in particular its behaviour and performance as a function of the undercut geometry and location on slabs-backing [16].

Installation procedures for dimension stone kerf anchorage are presented together with design procedures and core parameters affecting the anchorage strength [17].

Minimum slabs thickness formulae depending on the flexural strength capacity or pull-out strength capacity are provided taking into consideration stress concentrations either in the mid-span or in the support region for any of the abovementioned anchorage systems.

In some circumstances formulae to find maximum sag due to deformation under gravity load are presented when transverse permanent horizontal loads are supposed to induce long-term deformation.

Application examples to determine minimum thickness are presented for common and several slabs dimensions for all the referred anchorage systems.

Stainless steel and aluminium alloys are dealt exclusively as the main materials to use in anchors and supporting systems for stone cladding. Physical and mechanical properties of both metal alloys are provided in order to permit the designer to verify strength and deformation of the supporting elements and connections. Formula is presented in a unified way given the similarity of their ductile behaviour.

Design of members subjected to combined forces, such as bending and shear or shear and tension are illustrated making use of the dowel and pin, and the kerf anchor systems.

# References


1. Nacheman RJ (2006) The empire state building – façade evaluation and repair of an engineering landmark. Structure. C3 Ink, Wisconsin
2. Loughran P (2007) Failed stone. Birkhäuser, Boston
3. Hugues T, Steiger L, Weber J (2005) Dressed stone: types of stone, details, examples. Edition Detail
4. Lewis MD (1995) Modern stone cladding: design and installation of exterior dimension stone systems, vol MNL 21, ASTM manual series. ASTM, Philadelphia
5. Camposinhos RS, Lello JC (2010) Prestressed façade stone panels. In: Odasi E-TMM (ed) 7th International marble and natural stone congress of Turkey, Afyonkarahisar, Turkey, 2010. Ekim-2010 TMMOB Maden Muhendisleri Odasi, Ankara, pp 49–54
6. Camposinhos RS, Amaral P (2007) Caderno técnico de aplicação, uso e manutenção de rochas ornamentais. Assimagra, Lisboa
7. Rousseau MZ, Poirier GF, Brown WC (1998) Pressure equalization in rainscreen wall systems. Construction technology updates, vol 17. Institute for Research in Construction, Ottawa
8. ASTM – American Society for Testing Materials (2010) Standard guide for selection, design, and installation of dimension stone attachment systems, vol ASTM C1242 – 10. American Standard of Testing Materials, West Conshohocken
9. ASTM – American Society for Testing Materials (2010) Standard guide for selection of dimension stone for exterior use, vol ASTM C1528 – 10. American Standard of Testing Materials, West Conshohocken
10. AFNOR (2000) Travaux de bâtiment – Revêtements muraux attachés en pierre mince. Partie 1 : cahier des clauses techniques, vol NF P 65-202-1. CSTB, Paris
11. British Standards Institute (2010) Code of practice for the design and installation of natural stone cladding and lining. Traditional handset external cladding, vol BS 8298–2:2010. BSI, London
12. Beall C (1990) Selecting a safety factor for stone. Aberdeen Group, Boston
13. Camposinhos RS (2012) Dimension stone design – partial safety factors: a reliability based approach. Proc ICE Construct Mater 165(3):145–159. doi:10.1680/coma.10.00018
14. Faddy MJ, Wilson RJ, Winter GM (2003) The determination of the design strength of granite used as external cladding for buildings. Case studies in reliability and maintenance. Wiley, St. Lucia
15. Camposinhos RS, Camposinhos RPA (2009) Dimension-stone cladding design with dowel anchorage. Proc ICE Construct Mater 162(3):95–104. doi:10.1680/coma.2009.162.3.95
16. RdS C (2012) Undercut anchorage in dimension stone cladding. Proc ICE Construct Mater 3:161–175. doi:10.1680/coma.11.00050
17. Camposinhos RS, Camposinhos RPA (2012) Dimension stone design – kerf anchorage in limestone and marble. Proc ICE Construct Mater 165(3):161–175. doi:101.1680/coma.2009.162.3.95

# Chapter 2
# Natural Stone Characterization

**Abstract** Based on the literature a brief description and classification of rocks is presented throughout this chapter with a particular emphasis on some of the most well-known or famous rocks.

Indicative values for physical or mechanical characteristics for different rock's types are given. Due to the scattering on its values, tests are practically mandatory, whenever it's required to use natural stone as a building material even with no structural requirements, so that the main relevant standards are referred.

The use of natural stone as cladding material is discussed upon its physical and mechanical properties obtained in test characterization. Surfaces finish briefly are discussed considering inherent alterations on them.

## 2.1 Introduction

Natural Stone in façades has been the solution for all buildings façade solutions. In the earlier years of human civilization this was not a solution but a "natural" way out, that mankind had to use to obtain a shelter and a proper defence against enemies attacks. Later, the demand for more comfort and a way to express their supremacy castles and similar constructions entailed the art and creativity of men on the stone work.

In general, the term "rock" is a naturally formed solid aggregate of one or more minerals, and it is what stone is made from. "Dimension stone" normally is used to refer to rock, which has been dressed, neat or, better saying, engineered or used by man in construction of some sort, usually as discrete slabs or blocks. Furthermore, important is the fact that this dimension stone is no longer where they were formed as a part of a larger cohesive rock mass. Examples are the use as "ornamental" cladding for buildings or the construction of columns and lintels in certain buildings or even of "stone" as counter tops in kitchens.

R. de Sousa Camposinhos, *Stone Cladding Engineering*,
DOI 10.1007/978-94-007-6848-2_2, © Springer Science+Business Media Dordrecht 2014

## 2.2 Rock Types

In common linguistic use, "a stone" may refer to any small piece of rock that may or may not have been polished or weathered and that is lying on Earth's surface.

Rocks are usually classified by mineral and chemical composition, by the texture of the constituent particles and by the processes that formed them. Three types or groups are generally used: igneous, sedimentary, and metamorphic. They may be further classified according to particle size.

Igneous rocks are formed when molten magma cools and are divided into two main categories: plutonic or irruptive and volcanic or extrusive. Irruptive or intrusive rocks result when magma cools and crystallizes slowly within the Earth's crust, while volcanic or extrusive rocks result from magma reaching the surface either as lava or fragmental expel. Examples of plutonic rocks are granite, diorite, syenite, etc., and for extrusive rocks examples we have, basalt, rhyolite and pumice, among others.

For the metamorphic rocks, the one that stands out most is, undoubtedly, the marble. However, other metamorphic rocks are also used for this industry, such as quartzite, gneiss, slate and other schist rocks.

Examples of sedimentary rocks are limestone, sandstones, breccias, dolomite, and others less commonly used for cladding purposes. Despite what is stated in the industrial point of view, this type of classification is not at all used, since it implies a very specific technical and scientific knowledge.

A more practical classification has been used which is self-oriented to commercial aspects. In general, it is common to "classify" stone as "granite," "marble" and "limestone." The slate designation is normally used to rocks that present well defined schistosity planes.

The scattering of the values of rock's properties and their characteristics is generally very high so that there's the need to carry out several tests depending on the location, depth and the volume extracted or removed amount. However, some ornamental rocks have a very low scatter, which is a benefit for structural use in construction.

The most-used rocks classification depends mainly on the process of its formation. Igneous rocks derive directly from the solidification of magma. Sedimentary rocks are formed on the surface of the crust from other rocks or and pre-existing materials, or also by chemical precipitation. Metamorphic rocks have its origin from pre-existing ones, due to the action of the so-called metamorphism agents.

Petrography and mineralogical characterization of a rock are a ruled matter requiring not only and among other issues, the description of their mineral constituents, such as the shape, but also their textural relationships and occurred deformations or alterations. In the following sections, the main characteristic of three abovementioned groups of rocks is presented.

**Fig. 2.1** Coarse-grained porphyroid biotitic granite, of *yellowish* colour, characterized by the large size and abundance of the feldspar megacrystals, usually showing roughly defined contours

## *2.2.1 Igneous Rocks*

Igneous rocks are classified according to the depth at which the consolidation occurs. If this occurs at a great depth, the rocks are classified as plutonic, e.g., the granite. If, otherwise, solidification occurs near the earth's surface or crust, the resulting igneous rocks are said to be volcanic – this is the case, for example, of the basalt. The main factor that determines the texture of an igneous rock is the cooling rate. Plutonic rocks may have granular or porphyritic texture, indicating that magma has gone through a two-stage cooling process, while some volcanic rocks are porphyritic thus indicating that magma have crystallized below a volcano but erupted before completed crystallization, forcing the remaining lava to crystallize more rapidly with much smaller crystals. Generally igneous rock present crystalline grain interlocking and do not show any preferred orientation, even though some rocks may present a process-oriented formation.

Their mineral constituents, fundamental or additions, are either light coloured (felsic minerals) such as quartz, calc-alkaline or sodic feldspars, muscovite, etc., or dark (mafic minerals) such as biotite, pyroxene, amphibole, olivine, etc.

The various feldspars determine the diverse appearance and colour of the granites (Fig. 2.1), such as the bright red, reddish, pink, yellowish, whitish, etc.

Granites have in general high resistance to frost and thaw cycles. Its main physical and mechanical property values are: the coefficient of linear thermal expansion ranging between $10 \times 10^{-6}$/°C and $20 \times 10^{-6}$/°C, the density approximately equal to 2.7, the compressive strength, that may vary from 100 MPa to around 300 MPa, the flexure tensile strength, which is about 10–20 % of the compressive strength, depending on the type, and the elastic modulus that normally ranges between 20 and 60 GPa.

As stated above, the word “granite” is used in a variety of ways by different people. A simple definition or a more precise definition may be used by either architects or engineers and petrologists. The definition of granite expands wildly when used by people who sell decorative stone such as countertops, tile and building veneer.

Some of the most well-known granites in Europe for façades cladding are the Balmoral, the Kuru Grey and the Baltik Brown from Finland, the Blanco Cristal, the Silvestre and the Rosa Porrino from Spain, The Bohus from Sweden, the Clair du Tarn and the Lanhelin from France, the Gebhartdt's and the Neuhauser from Austria, Pedras Salgadas and Cinza Alpalhão from Portugal.

Other igneous rocks have to be mentioned here, such as the Basaltina tephrite volcanic and the Trientiner Porphiry rhyolite from Italy, the Blue pearl from syenite from Norway, the Cinzento Grey foyaite from Portugal or the Wolga Blue anorthosihe from Ukraine.

There are publications that present the most granite dimension stones on the Internet which qualities are described in tables from different points of view, such as the commercial name, country, petrographic type, geological age, water absorption, similar stones, and basis of name.

### 2.2.2 *Sedimentary Rocks*

Sedimentary rocks are formed by the deposition of material at the Earth's surface and within bodies of water. Sedimentation gathers together the processes that cause mineral and organic particles – sediments – to settle and accumulate or minerals to precipitate from a solution. Before being deposited, sediment was formed by weathering and erosion in a source area and then transported to the place of deposition by water, wind, ice, mass movement or glaciers, which are called agents of denudation.

The sedimentary rock cover of the continents of the Earth's crust is extensive, but the total contribution of sedimentary rocks is estimated to be only 8 % of the total volume of the crust. In fact, they are only a thin covering over the Earth's crust, consisting mainly of igneous and metamorphic rocks. Based on the processes responsible for their formation, sedimentary rocks can be subdivided into three main groups: clastic, biochemical and chemical sedimentary rocks.

Clastic or “mechanical” sedimentary rocks are composed of silicate minerals and rock fragments that were transported by moving fluids and were deposited when these fluids came to rest. Clastic rocks are composed largely of quartz, feldspar, rock fragments, clay minerals, and mica; other numerous minerals may be present as accessories as in Fig. 2.2.

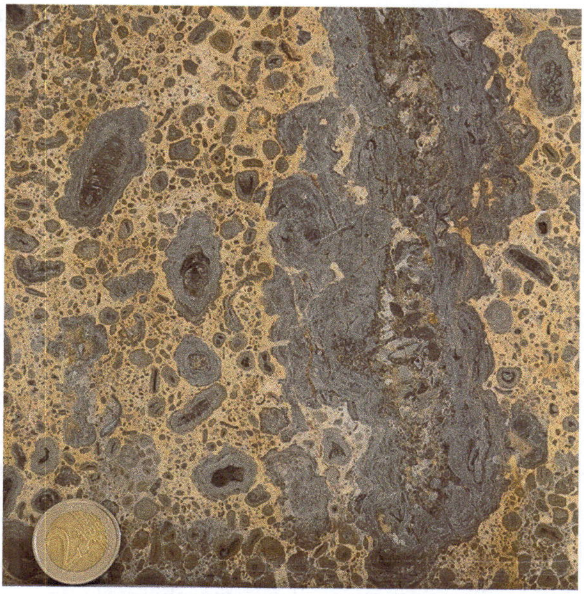

**Fig. 2.2** A *golden* and *greyish* limestone with *reddish* spots, of breccious appearance, bioclastic, sometimes calciclastic, with oncolites

Biochemical sedimentary rocks are created when organisms use materials dissolved in air or water to build their tissue. Examples include most types of limestone formed from the calcareous skeletons of organisms such as corals, and molluscs. Deposits of chert, formed from the accumulation of siliceous skeletons from microscopic organisms, are also examples of the formation of this subtype of sedimentary rocks.

Chemical sedimentary rock forms when mineral constituents in solution become supersaturated and inorganically precipitate. Common chemical sedimentary rocks include oolitic limestone and rocks, composed of water-soluble mineral sediments, that result from concentration and crystallization by evaporation from an aqueous solution.

As one of the most exploited sedimentary rock limestone presents considerable resistance to frost and thaw cycles. Limestone in a commercial sense is less dense, with higher water absorption than granite or even marble. They are usually used for fewer graded purposes, as building stones, or honed limestone flooring.

The main characteristics are: density, which is approximately equal to 1.8, the average compressive strength between 50 and 60 MPa, the flexure tensile strength which is about 5 MPa, and the coefficient of linear thermal expansion ranges between $4.0 \times 10^{-6}$/°C and $8 \times 10^{-6}$/°C. The average values of the modulus of elasticity lay between 10 and 80 GPa.

Belgian Red, Belgishdch "Granite" from Belgium, Adneter from Austria, Travertino Romano, Nero Portoro or Botticino from Italy, Rojo Alicante from Spain, Savonnieieres from France, Estremadura Azul from Portugal are some of the Limestone or travertine well-known rocks in Europe, not to mention other famous sedimentary rocks like the Warthauer and the Sirkwitz-Rackwitzer from Poland, Bollinger and Rorschache from Switzerland, Nexö from Denmark, all of them sandstones used with construction purposes.

### 2.2.3 Metamorphic Rocks

Metamorphic rocks arise from the transformation, i.e., the change in form or metamorphism of existing rock types. The original rock is subjected to heat and pressure causing profound physical and chemical change. The protolith, or original rock, may be sedimentary rock, igneous rock or another older metamorphic rock.

Metamorphic rocks make up a large part of the Earth's crust and are classified by texture and by chemical and mineral assemblage. They may be formed simply by being deep beneath the Earth's surface, subjected to high temperatures and the great pressure of the rock layers above it. They can form from tectonic processes such as continental collisions, which cause horizontal pressure, friction and distortion. They are also formed when rock is heated up by the intrusion of magma from the Earth's interior. Some examples of metamorphic rocks are gneiss, slate, marble, schist, and quartzite.

In nature, it's possible to find two very different kinds of metamorphic rocks: those which resulted from contact metamorphism and those which resulted from regional metamorphism.

For example, marble is formed trough contact metamorphism, on the other side, schistosity is one of the most common structures metamorphic rocks subject to regional metamorphism and refers to the parallel orientation of the mineral constituents, elongated or tabular. One of ornamental rocks in which this aspect is well marked is the slate. The foliation is also another typical structure of the rocks that have undergone regional metamorphism.

A marble is a metamorphosed limestone. It is a crystalline rock composed of calcite, or, more rarely, dolomites. The microscopic calcite grains recrystallize to form macroscopic crystals presenting a sacaroid aspect. The main texture is granoblastic with equidimensional grains. As well as limestone, marble is characterized by low hardness and effervescence with cold hydrochloric acid, when formed by calcite, and hot, when formed by dolomite. The marble generally is white, but may have a wide range of colour caused by other minerals in small amounts (Fig. 2.3).

These minerals, arising from impurities present in the original lime are mainly mica, chlorite, graphite, and serpentine, among others.

The term "marble" is often used in a commercial sense meaning any rock composed of calcium carbonate, which can be polished. By this, some limestone rock types are also labelled "marbles."

Indeed, limestone should not be confused with marbles while the former are generally sedimentary rocks consisting of aggregated clastic elements trough calcitic cement, marble is derived from the action of metamorphic phenomena on

**Fig. 2.3** A fine grained *white* marble with *greyish* and *green* small veins

the limestone consisting mainly of authigenic calcite. This distinction is reflected in a marked difference of the physical and mechanical characteristics of both types of rock.

Estremoz from Portugal, Dionysos from Greece, Rusita from Romania and Carrara from Italy are one of the most famous marbles in Europe even though other metamorphic rocks have to be mentioned such as the Iragana or the Calanca paragneiss from Switzerland the Alta from Norway and the Verde Spluga from Italy, all famous fine-grained quartzite.

Marble in a commercial meaning is a softer natural stone than granite, and more dense than limestone.

The marble main physical and mechanical values are the density of about 2.6, the average compressive strength between 50 and 90 MPa; the flexure tensile strength which is about 7 MPa and the coefficient of linear thermal expansion, which has a value between $7 \times 10^{-6}/°C$ and $22 \times 10^{-6}/°C$. The average values of modulus of elasticity ranges between 40 and 80 GPa.

## 2.3 Natural Stone Standards for Cladding

Information that can help stone specialists, sellers, engineers and architects to orient them in a thoughtless situation of stone commercial names and commercial and scientific stone types may be found at the Internet. However, in what to safety requirements and suitability to use concerns this information is, at least, scarce.

Without an unwavering, realistic set of standards and testing procedures for stone products, the stone industry as a whole would be in confusion. The standards that have been developed and set in place for these products are important tools to help protect end users, individual companies and the industry from unconformities and product failures. Material standards help to shun the use of stone products for unsuitable applications.

In this context physical and mechanical characterization of natural stone is required as well as laboratory tests to assess its aptness to the various conditions of use, a procedure that's essential and decisive when dimension stone cladding design and construction are the key issue.

Several methods for testing and comparing the physical properties of natural building stones have been developed. The most basic stone testing methods analyze three characteristics: the density, water absorption and the compressive strength. Although these three tests are typically run as a suite with modulus of rupture and flexural strength, for practical purposes, flexural strength tests are the most relevant to façade stone cladding methods once structural problems are determinant.

Most architectural specifications require that stone meets certain specified, European Committee for Standardization (CEN), American Society for Testing Materials (ASTM) or other testing standards before it is accepted for use. For instance in the European market, the harmonized standards and CE marking have important consequences for producers. It is the responsibility of the producer, or, more correctly, the supplier of the final natural stone product, to ensure that the properties of the product are documented before the product is sold or being used as a construction material. It must be said that in some European countries it is against the law, to sell a product without such documentation, since the period allowed for national implementation of the standard is over.

It must be emphasized that the major innovation in the European standards in contrast to the traditional national standards is that they are geared to evaluation of conformity and factory production control. It is now required that the supplier or manufacturer of dimension stone products shall be responsible for the assessment and attestation that the products offered are in conformity to the requirements of European standards. The compliance with the requirements of the published standards and stated values, initial type testing and permanent production control, with results registration have to be made available.

In Canada and United States, it's generally required that dimension stone comply with specified ASTM or other testing standards before it is accepted for use.

For example, the determination of flexural strength under concentrated load or under constant moment is mandatory, and the coefficient of the variation of this property must be provided for practitioners and designers of dimension stone.

Tests can be divided into two categories: those, including the identification and characterization of physical properties and those, related with the mechanical characterization.

The former aims to identify and determine the fundamental characteristics of natural stone at a macro and microscopic scale, as well as to gather their main physical properties.

There are several standard test methods to evaluate stone characteristics so that stones can be compared on a uniform basis. These methods and standards for dimension stone are the basis to establish guidelines for specification and installation. American Society for Testing and Materials (ASTM) International and the European Committee for Standardisation (CEN) are the world's leading standards development organizations. It has to be noted that stone testing according to European methods and conditions may use different procedures that give different results than ASTM methods do for the same stone. This is particularly true for tests for abrasion.

In the following lines, the more relevant standards are presented for cladding issues, regardless those revetments such as internal or external paving, kerbs or others like purpose's fittings when for example, abrasion or slip resistance is crucial.

For the description and physical characterization the more relevant standards and test methods provided by the CEN Technical Committee 246 are as listed:

- EN 12057 – Natural stone products. Modular tiles. Requirements;
- EN 12370 – Determination of resistance to salt crystallization;
- EN 12371 – Determination of frost resistance;
- EN 12407 – Petrographic examination;
- EN 12440 – Natural stone. Denomination criteria;
- EN 12670 – Standard Terminology relating to dimension stone;
- EN 13373 – Determination of geometric characteristics on units;
- EN 13755 – Determination of water absorption at atmospheric pressure;
- EN 14066 – Determination of resistance to ageing by thermal shock;
- EN 14581 – Determination of linear thermal expansion coefficient;
- EN 1469 – Natural stone products — Slabs for cladding — Requirements;
- EN 1925 – Determination of water absorption coefficient by capillarity;
- EN 1936 – Determination of real and apparent density and of total and open porosity.

The significant standards and test methods provided by the CEN Technical Committee 246 for the Mechanical characterizations are:

- EN 12372 – Determination of flexural strength under concentrated load;
- EN 13161 – Determination of flexural strength under constant moment;
- EN 13364 – Determination of the breaking load at dowell hole;
- EN 14146 – Determination of the dynamic elastic modulus of elasticity;
- EN 14580 – Determination of static elastic modulus;
- EN 1926 – Determination of uniaxial compressive strength.

In the same manner, the ASTM International has also developed similar standards and test methods to evaluate stone characteristics so that stone can be compared on an intrinsically uniform basis, yet the engineer or practitioner must be aware of the differences on the numerical results that he certainly should find. For the terminology and the physical characterization one has:

- C 119 – Standard Terminology Relating to Dimension Stone;
- C 1242 – Standard Guide for Selection, Design, and Installation of Dimension Stone Attachment Systems;
- C 1528 – Standard Guide for Selection of Dimension Stone;
- C 1721 – Standard Guide for Petrographic Examination of Dimension Stone;
- C 97 – Standard Test Methods for Absorption and Bulk Specific Gravity of Dimension Stone;
- D 2203 – Standard Test Method for Staining from Sealant;
- D 4404 – Standard Test Method for Determination of Pore Volume and Pore Volume Distribution of Soil and Rock by Mercury Intrusion Porosimetry;
- D 4341 – Standard Test Method for Creep of Cylindrical Hard Rock Core Specimens in Uniaxial Compression;
- D 5312 – Standard Test Method for Evaluation of Durability of Rock for Erosion Control under Freezing and Thawing Conditions.

In relation with mechanical characterization the relevant ASTM standards are:

- C 1201M – Standard Test Method for Structural Performance of Exterior Dimension Stone Cladding Systems by Uniform Static Air Pressure Difference;
- C 120M – Standard Test Methods of Flexure Testing of Slate (Breaking Load, Modulus of Rupture, Modulus of Elasticity);
- C 1352M – Standard Test Method for Flexural Modulus of Elasticity of Dimension Stone;
- C 1354M – Standard Test Method for Strength of Individual Stone Anchorages in Dimension Stone;
- C 170M – Standard Test Method for Compressive Strength of Dimension Stone;
- C 503 – C568 – C615 – Standard Specifications for Marble; Limestone; Granite Dimension Stone;
- C 880M – Standard Test Method for Flexural Strength of Dimension Stone;
- C 99 – Standard Test Method for Modulus of Rupture of Dimension Stone;
- D 4341 - 93 Standard Test Method for Creep of Cylindrical Hard Rock Core Specimens in Uniaxial Compression;
- D 7070 Standard Test Methods for Creep of Rock Core under Constant Stress and Temperature.

## 2.4 Natural Stone Properties

As already mentioned, stones given commercial names can lead to confusion even if the traditional name for the stone is well-known and suitable historical information can be found. The wording and terminology should reflect the requirements of the relevant standards. A macroscopic description is possible on any available sample

**Table 2.1** Physical and mechanical properties indicative values

| Group | Rock type | Apparent density mass (Kg/m$^3$) | Water absorption at atmospheric pressure (mass %) | Uniaxial compressive strength (MPa) | Thermal expansion coefficient (mm/mK) |
|---|---|---|---|---|---|
| Igneous | Granite | 2,600–2,800 | 0.1–0.9 | 130–270 | 0.008 |
| | Syenite | 2,600–2,800 | 0.2–0.9 | 160–240 | 0.008 |
| | Diorite | 2,800–3,000 | 0.2–0.4 | 170–300 | 0.009 |
| | Gabbro | 2,800–3,000 | 0.2–0.4 | 170–300 | 0.009 |
| | Periodite | 2,500–2,800 | 0.2–0.7 | 180–300 | 0.013 |
| | Basalt | 2,900–3,000 | 0.1–0.3 | 240–400 | 0.1–0.3 |
| | Diabase | 2,800–2,900 | 0.1–04 | 180–250 | 0.1–0.4 |
| Sedimentary | Breccia | 2,600–2,750 | 0.1–1.0 | 50–160 | 0.003 |
| | Conglomerate | 2,200–2,500 | 0.8–1.0 | 20–160 | 0.002–0.003 |
| | Sandstone | 2,000–2,700 | 0.2–10.0 | 30–150 | 0.012 |
| | Limestone | 2,600–2,900 | 0.1–3.0 | 75–240 | 0.004–0.001 |
| | Dolomite | 2,600–2,900 | 0.1–3.0 | 75–240 | 0.005 |
| | Travertine | 2,400–2,500 | 2.0–5.0 | 20–60 | 0.007 |
| Metamorphic | Marble | 2,600–2,900 | 0.1–3.0 | 75–240 | 0.003–0.006 |
| | Migmatite | 2,600–3,000 | 0.3–0.4 | 100–200 | 0.005–0.008 |
| | Slate | 2,600–2,800 | 0.2–0.4 | 140–200 | 0.005–0.001 |
| | Quartzite | 2,600–2,700 | 0.2–0.5 | 150–300 | 0.0125 |
| | Gneiss | 2,600–3,000 | 0.3–0.45 | 100–200 | 0.005–0.008 |

that shows the general colour, tone and texture of the stone, but cannot show the range of geological characteristics that will be naturally present in a given type in order to assess its characteristics.

Nevertheless, it's useful to have some idea about the range for the values to be expected in the most of the rocks. From the tests that are usually performed to obtain the physical characteristics of rocks, those which stand out for design purposes are regularly used for any characterization study. In general, it is reasonable to assume that a particular rock can be identified by its colour, texture, mineralogical composition and by its physical and mechanical properties.

The values and its range given in Table 2.1 may only be used as an indication in rock identifying. They were obtained from specific literature according to European CEN standards.

In this point petrography analysis is essential for architects and engineers in determining how the stone will behave or has been behaving in service. Following some appropriated standard and after having examined prepared samples using a petrographic microscope, the petrographer may determine the proper scientific classification and investigate characteristics important for the performance of the stone.

Terminology is also important to be taken in a precise way thus avoiding misinterpretations and to facilitating the understanding and transmission of information on the geometrical characteristics of the products. For example, EN 12670 defines a plate or cladding slab as an element with two dimensions, predominantly greater than the third called "thickness."

**Table 2.2** Physical and mechanical properties indicative values

| Nominal thickness (t) (mm) | Tolerances |
|---|---|
| $30 \geq t > 12$ | ±10 % |
| $80 \geq t > 30$ | ±3 mm |
| $t > 80$ | ±5 mm |

**Table 2.3** Length tolerances according to EN 13373

| | Length and or width (L) [mm] | |
|---|---|---|
| Thickness [mm] | L < 600 | L ≥ 600 |
| $t \leq 50$ | ±1.0 mm | ±1.5 mm |
| $t > 50$ | ±2.0 mm | ±3.0 mm |

### *2.4.1 Geometry and Tolerances*

The thickness of a dimension stone slab is attained after sawing or other precise and delicate techniques, such as diamond wire saws, diamond belt saws, burners and surface finishing. For cladding purposes their final dimensions may depend on the fixing method to the backup, either by mechanical means, *i.e.*, using anchorage devices, or by cement bonding using adhesives, cement, epoxy resin or other type compound adhesive material.

Length, width and thickness of a dimension stone shall be defined in this order and according to the referred standards, e.g., EN 13373, unit sizes should be obtained within the range of tolerance that is depicted in Table 2.2.

This European standard describes methods for verifying the geometric characteristics of products of natural stone such as blocks, rough slabs and finished products for cladding, flooring, stairs and tiles. These methods are to be applied for the case as a dispute between two parties; they are not compulsory for production control.

Whether the slabs thickness is greater or smaller than 50 mm, tolerances for length and width are also specified in EN 13373, depending on the size dimension according to Table 2.3.

In the same manner holes, cuts or kerfs size variations are also limited. For instances, the specific position, diameter and depth dowell hole sizes have to be performed with a tolerance of 2 mm in the slab's thickness, its depth and diameter have to lie in a range of (+3; −1 mm) and (+1; −0.5 mm) respectively.

Flatness and squareness are also limited. For not calibrated modular tiles the corresponding value is 0.15 %. In a calibrated tiling process the limit to those properties is 0.10 % according EN 12057.

### *2.4.2 Density, Porosity and Water Absorption*

Natural stone's density is very sensitive to the minerals that compose a particular rock type. Those rocks which are rich in quartz and feldspar tend to be less dense than volcanic rocks; the more mafic a rock is, the greater its density. The density of a material such as a rock has to be clearly defined: the bulk density, which is controlled by the porosity and the degree of cementation, and the matrix density, that depends on the components and does not take the porosity into account.

After knowing the weight of a sample under dry conditions, $m_{dry}$, underfully saturated conditions, $m_{sat}$, and the sample weights submerged in water, $m_{sub}$, the matrix density, $\rho_{matr}$, and the bulk density, $\rho_{bulk}$, may be determined using the following equations:

$$\rho_{matr} = \rho_{water} \cdot (1 - m_{sub}) \tag{2.1}$$

The bulk density is obtained as follows:

$$\rho_{bulk} = \rho_{water} \cdot \left( \frac{m_{dry}}{m_{sat}} - m_{sub} \right) \tag{2.2}$$

In general the bulk density is said to be high when its value exceeds 2,800 kg/m$^3$ and it's usually considered low when that value is less than 2,300 kg/m$^3$.

The fraction of void space in a material is defined as the total porosity, $p_{tot}$, of it and may be determined dividing the volume of void space by the bulk volume of a given specimen or sample.

Alternatively the total porosity can be calculated from the bulk density and matrix density as follows:

$$p_{tot} = 1 - \frac{\rho_{bulk}}{\rho_{matr}} \tag{2.3}$$

The effective or open porosity in spite of the former does not comprise all pore spaces in a sample but only voids where fluids and air can access. The open porosity, $p_o$, may be obtained using the following equation:

$$p_o = \left( \frac{m_{sat} - m_{dry}}{m_{sat} - m_{sub}} \right) \times 100 \tag{2.4}$$

Water absorption is a measure of the effective porosity of a stone and can be an indicator of its susceptibility to damage during freezing. A stone that has greater water absorption will also tend to absorb liquid stains more readily. In general, the lowest water absorption is desired.

The water absorption at atmospheric pressure, $W_a$, is the ratio between the water absorbed by a sample and its dry weight, a percentage value of its density, and may be obtained as follows:

$$W_a = \frac{m_{sat} - m_{dry}}{m_{dry}} \times 100 \tag{2.5}$$

Different values may be found depending on the rock type, nevertheless, granites rarely present water absorption values higher than 1 %, for marbles and limestone the value of 3 % is not usual and for Schist, 0.5 %, is a high value also [1, 2].

### 2.4.3 *Coefficient of Thermal Expansion*

Continuous changes in temperature occur in façades, veneers or any other external surfaces, due to the solar incidence and/or air temperature. The thermal amplitude value, depending on the stone's colour, the exposure conditions and the climate, can reach even in the same period of 24 h near 50 °C [3].

These temperature ranges in the material cause considerable dimensional variations in any solid element as is the case of natural stone. When the corresponding movements, either stretching or shortening, are restricted or not considered in the design, stress variations will arise in the interface layers or contact points of different materials. These stresses may result in considerable damage, giving rise to several well-known pathologies and sometimes breakage of failure with cladding's detachment. It must be emphasized that pathologies in natural stone claddings generally exhibit different anomalies that are caused by several phenomena either than those induced by thermal stresses.

The coefficient of thermal expansion describes how the size of an object changes with a change in temperature. The linear coefficient is generally used for natural stone characterization and in some cases it may be necessary do joint's design when one might only be concerned with the change along a length, or over some area.

One of the most referred problems related with the thermal expansion is the temperature-induced damage even at low temperature intervals is the bowing of plates [4], a phenomena that is currently observed in thermally sensitive stones such as calcitic marbles, which shows residual strain after heating.

### 2.4.4 *Weathering Resistance*

Cladding dimension stones have small size thickness when compared with their length and width. Even though they have no structural purposes, mechanical requirements have to be considered on account of safety reasons due to the imposed flexural and anchoring stresses and also to the weathering effects.

When exposed to the atmosphere, natural stone experiments a slow continuous process depending on the exposure conditions. Either mechanical or chemical decay on its characteristics is to be observed due to several factors. Freeze and thaw or wetting-drying cycles, variation in temperature, mechanical induced actions, and salt crystallization in pore spaces are some of the most common agents, that cause stone weathering [5].

Salt crystallization and freeze/thaw tests have to be made in order to know how weathering depends on material characteristics such as pore space, water transportation and mechanical features. In the same manner, resistance to ageing by thermal shock is also necessary to verify the decay of natural stone properties to be used as a construction material.

In the following sections, the assumptions that underlie the pivotal aspects in what the weathering due to harsh environment concern are discussed.

#### 2.4.4.1 Frost Resistance

Frost resistance of natural stones is multifarious phenomena. It depends on the rock's properties and on the presence of salts, temperature variation, humidity, etc. In some climates frost may be considered as one of the most important factors that cause natural stone decay properties. Marbles and limestone as well, due to its intrinsic properties, porosity and mineral composition, are in general less resistant than granites.

By freezing, water increases in volume by about 10 % and that's why the formation of ice inside the rock's structure gives rise to tensile stresses. Therefore, due to its reduced tensile strength, the internal stresses are potentially failure-inducing, particularly in high porosity natural stones such as dolomite and sandstone.

The bigger part of the damage observed on natural stone façade claddings due to frost is in sandstone and certain types of limestone, and is has been proved that low salt concentrations at the surface of the stone severely magnify the effect of frost attack on some types of stones [6]. The increase in air pollution instigates salts deposition on the surface of buildings façades, thus explaining why stone damage had augmented during the last decades and made the stone more sensitive to surface scaling from frost.

Standardized tests are aimed to measure the durability of the material in environments subject to freezing and thawing. It involves putting a series of samples through freeze/thaw cycles, from, −12 to +20 °C. After each cycle, the samples are first visually inspected to check for cracks, deterioration or loss of fragments, and then given a flexural strength test.

In general, it may be considered as frost-proof a stone that does not change the measured flexural strength before and after frost testing less than or equal to 20 %.

The number of cycles, $N$, to be required has to depend on the climate and on the cladding's surface condition and orientation. González-Mesones [7] proposed a formula to estimate its value depending on the region by mean of an index, $I_g$, the reference minor test temperature, $t$, −12 °C in the case of the CEN, the building life cycle, $n$ and a saturation factor, $k$, reflecting the exposure conditions which is related with the surface's cladding situation in a building.

The ice index, $I_g$, represents, for a given region or geographic zone, the annual average value of the sum of the air temperatures, $T$, lower than −5 °C over a period of 30 years, daily measured. Its value is thus calculated according to the following formula:

$$I_g = -\frac{\sum_{0}^{365\times30} T_{(\leq -5)}}{30} \tag{2.6}$$

The proposed acceptance criteria for the minimum number of satisfactory test cycles are given by:

$$N \geq \frac{I_g \cdot n}{|t|} \cdot \mathrm{k} \tag{2.7}$$

For mortar based fixing claddings and for ventilated façades, the given value of, k, is 0.4 and 0.05 correspondingly.

It must be emphasized that a conservative procedure might be to consider as frost-proof stones that demonstrate flexure strength decay, after 48 test cycles, lesser than 20 %, which is the case of almost all intrusive rocks such as granite.

Certain indirect methods of stone durability assessment that investigated the pore properties of the stone were found to be also a trustworthy predictor of frost resistance. Petrographic examination seems to be a valuable way for observation of changes caused by frost action [8].

#### 2.4.4.2 Resistance to Salt Crystallization

Rainwater in contact with limestone, on evaporation, makes any calcium carbonate dissolution precipitate in the form of calcite crystals. Salt is also an important weathering factor as a consequence of exposure to industrial polluted environments in coastal areas and in places where salt is used as antifreeze. In fact, the crystallization of salt even in unsaturated environments is often accompanied by a volume increase, causing internal pressures that result in irreversible damage. Either by cracking, exfoliating or disintegrating of macrostructure a loss of mass and strength is observed [4].

Tests in general are aimed to quantify durability against salt crystallization weathering action. These are usually based on the immersion of samples into a saline solution. Regarding standard EN 12370, as in the frost resistance test, samples are subjected to cycles of 24 h. Each cycle consists of soaking in saline solution of sodium sulfate (2 h), followed by drying in an oven (20 h) and further cooling (2 h). A total of 15 cycles is required.

The results are expressed as a percentage of weight decay, $\Delta M$, to the dry initial weight. Even though there are several proposed criteria to define stone suitability either by ultrasonic velocity measurements or compressive strength measurements before and after salt exposure [9], the Building Research Establishment (BRE) criteria based upon the loss of mass [10] has been widely used. It consists of the definition of acceptance limit values of mass loss depending on the exposure and environmental conditions. Six categories (A to F) are proposed depending on the mass loss values.

The harshest conditions such as in coastal areas near urban or industrial areas class B are pointed out as the lowest admissible. In intertidal zones in non-occupied areas a class C or higher are acceptable whether class D is the required minimum category for interior regions where pollution may exist. The class definition is obtained according to Table 2.4.

Another particular factor to take into account is that in the bottommost parts of the building façade, closest to the soil, a saline solution can ascend through the stone by capillarity and the inherent evaporation with consequent crystallization of salts can occur. Contact between the stone and the soil must then be avoided.

**Table 2.4** Durability classes for salt crystallization resistance according the Building Research Establishment

| Resistance class | Mass loss |
|---|---|
| A | $1\,\% \leq \Delta M$ |
| B | $1\,\% \leq \Delta M < 5\,\%$ |
| C | $5\,\% \leq \Delta M < 15\,\%$ |
| D | $15\,\% \leq \Delta M < 35\,\%$ |
| E | $\Delta M \geq 35\,\%$ |
| F | Fracture or cracking before 15 cycles |

#### 2.4.4.3 Resistance to Ageing by Thermal Shock

As it has been mentioned above the temperature changes may attain considerable values inside the stone volume, and the continuous heating and cooling of the stone lead to its deterioration over the time due to the corresponding dimensional changes.

The criteria generally adopted to verify stone suitability consists of observing the decreasing mass and of the elastic modulus of natural stone produced by abrupt changes in temperature. For this, the samples are subjected to specific cycles, as the tests of resistance, to freezing and the salt crystallization. According to EN 14066, each cycle comprises a period of drying in an oven at 105 °C for 18 h, followed by a period of water cooling at 20 °C for 6 h.

After 20 test cycles, the mass decrease is determined in relation to the initial mass of the sample as well as that of the dynamic elastic modulus value.

The observed decrease value being superior than 20 %, the stone shall be rejected for areas predominately sun oriented in hot climates.

Temperature and moisture variations are important factors in stone degradation. For some distinct reasons these variations cause volume changes. For example, in granite or some sandstone the different expansion coefficients of the minerals present on the stone will result in volume expansion or contraction. Due to other reasons, calcitic marbles exhibit granular decohesion once calcite expands in one direction and contracts in the other. The well-known phenomenon of the so called "sugaring" deterioration is directly dependent on these uneven deformations.

The dolomite marbles should not be so prone to thermal shock once expansion or contraction occurs in the same direction, though dolomite marbles are also susceptible to thermal cyclic due to variations in its texture that affects the distribution of thermal stresses [11].

### *2.4.5 Mechanical Characterization*

The differentiation between the physical and mechanical properties of a solid material, such as stone, may be ambiguous, yet properties such as strength, Young's Modulus, creep, Poisson's ratio, fracture toughness, among others, are clearly defined as mechanical properties.

An acceptable definition for the mechanical characterization of a given material would be the quantitative identification of the material's behaviour, *i.e.*, how it reacts to physical forces. In fact, mechanical properties of a material occur as a result of its physical properties for which the values may be given through a series of standardized tests. These tests are aimed to determine the fundamental characteristics of natural stones in mechanical terms. Examples are the compressive strength, tensile strength, flexural strength, etc.

Both the ASTM Committee C18 and the CEN Technical Committee 246 has jurisdiction of several standards on mechanical characterization of dimension stone.

The following standards play a preeminent role in what building's envelope mechanical requirements concerns.

- ASTM C1201/C1201M – Standard Test Method for Structural Performance of Exterior Dimension Stone Cladding Systems by Uniform Static Air Pressure Difference;
- ASTM C1352/C1352M – Standard Test Method for Flexural Modulus of Elasticity of Dimension Stone;
- ASTM C1354/C1354M – Standard Test Method for Strength of Individual Stone Anchorages in Dimension Stone;
- ASTM C170/C170M – Standard Test Method for Compressive Strength of Dimension Stone;
- ASTM C880/C880M – Standard Test Method for Flexural Strength of Dimension Stone;
- ASTM C99/C99M – Standard Test Method for Modulus of Rupture of Dimension Stone;
- EN 12372 – Natural stone test methods – Determination of flexural strength under concentrated load;
- EN 13161 – Natural stone test methods – Determination of flexural strength under constant moment;
- EN 13364 – Natural stone test methods – Determination of the breaking load at dowell hole;
- EN 14146 – Natural stone test methods – Determination of the dynamic modulus of elasticity (by measuring the fundamental resonance frequency).
- EN 14580 – Natural stone test methods – Determination of static elastic modulus;
- EN 1926 – Natural stone test methods. Determination of uniaxial compressive strength.

The strength of stone depends on the rock structure, and it is influenced not only by its composition, *i.e.*, shape and size of its constituents, but also by the aging and exposure conditions. It is well-known that stones with high specific weight and low water absorption have greater strength and elasticity module. Because the structure and cementing of the grains of some stone types allows for a bigger adsorption of moisture, decay in compressive or flexural strength is a result of those stones being saturated with water even at atmospheric pressure [12, 13].

Together with strength loss, bowing is also known to occur in some marbles, when used as exterior claddings. Quantitative test methods as an evaluation of marble being able to resist bowing and strength induced by thermal and moisture, are under development by the ASTM Committee C18 and the CEN Technical Committee 246. Relating this topic the following pre-standards have to be mentioned:

prEN – 16306 Natural stone test methods – Determination of resistance of marble to thermal and moisture cycles;
ASTM – WK32888 – New Test Method for Determination of the resistance of marble to thermal and moisture cycling.

The samples to be tested should be drawn from normal production blocks, extracted from the same quarry defined for a given project. This is to say, that historical data only gives an indication on strength in an informational relative basis about the stone to be used.

Designers must be aware that stone is anisotropic. Its properties are directionally dependent.

The method of deposition in sedimentary rocks, in which the layers or beds are predominantly orientated or bedded, or in the case of metaphoric rocks, sedimentary transformed by heat and pressure or, even in the case of igneous rocks, due to the cooling of the structure, which may cause micro predominant cracking in a particular direction or directions. These formation phenomenona gave rise to anisotropic behaviour.

The stone appearance in various directions will ask for a decision on the preferred orientation. Thus, the aesthetical effect may have a strong influence on the mechanical relevant properties and then on the cost. This is a key issue to consider during initial testing.

In the following sections, the significant mechanic properties and tests are addressed, regarding dimension stone cladding and veneers design.

#### 2.4.5.1 Compressive Strength

Even though the compressive strength is not determinant for dimension stone cladding design, this property may be regarded as a dominant feature of any stone. Several other properties are consistent with compressive strength. Designer should always compare relative values of other properties such as flexural strength, elasticity modulus or specific weight. Furthermore, minimum strength requirements might be specified according to the standards.

Samples may differ in sizes and shape: cut cubes or on core-drilled cylinders are subject to uniaxial loading in a calibrated test machine until rupture is obtained. The maximum applied load is divided by the contact area of the specimen to calculate the compressive strength. Usually 5–10 samples in dry or soak conditions are used. The load is evenly applied perpendicularly or parallely to the anisotropy planes at a fixed speed by a load cell.

#### 2.4.5.2 Flexural Strength

Based on the formula for linearly elastic bodies, the flexural strength is defined as the maximum tensile stress in a rock specimen when it is about to fracture. Prismatic specimens are obtained from the same larger panel or slab.

Three or four point bending flexural tests are performed on each specimen, the former by applying a single load midspan, the latter with loads at the quarter points of the stone specimen.

The load is increased gradually until the test specimen fractures. The maximum applied load is recorded and the flexural stress that occurred in the specimen at fracture is calculated.

Several dimensions are possible for the prismatic specimens; the thickness shall be between 25 mm and 100 mm and shall be greater than twice the size of the largest grain in the stone; the total length shall be equal to six to ten times the thickness depending on the standards. The width shall be between 50 mm and three times the thickness and it shall be less than the thickness.

Specimen's face in tension is to have a fine abrasive, sawn, honed or polished finish, yet it will need that surface finishes, *e.g.*, flamed, sandblasted as required for application, are be tested accordingly.

Depending on the visual orientation and directional properties, care should be taken to gather the critical direction of testing. It should be noted that to fully understand the micro-structure of the material testing in three orientations may be necessary initially. The same way, if wet testing results present lower values than dry testing, then dry testing results have to be overlooked for design purposes.

For different reasons it's expected to obtain lower values from four-point flexure tests than from three-point flexure tests. One of the major reasons for the discrepancy depends on the fraction of the specimen's volume which undergoes the greatest tensile stress. This is the basis of the so-called weakest-link theory which elucidates the influence of the scale effect. Its principles were enunciated by Weibull and Burshtein [14, 15].

#### 2.4.5.3 Tensile Strength

Tensile strength is formally defined as the tensile stress required to cause the failure of an unconfined cylindrical or cubical stone specimen, divided by the cross-sectional area of the specimen perpendicular to the axis of loading. This is the direct tensile strength; because of the difficulties related to gripping the specimens, this is a very unusual test [16].

Otherwise, the tensile strength can be found indirectly, i.e. by relying on another type of test. One of these indirect tensile strength methods is the so-called Brazilian test, where a circular solid disc is compressed until failure across a diameter; tensile stresses perpendicular to that diameter plane are developed; as such, compressive loading machines are used in this test. In the Brazilian test a stone's indirect tensile strength, $\sigma_{Rt}$, is generally defined as:

$$\sigma_{Rt} = -\frac{F}{\pi \cdot r \cdot t} \tag{2.8}$$

with $F$ the compressive force, $r$ the radius of the specimen (disc) and, $t$, the thickness of the disc.

Yet, it must be said, that the above formula for determining the indirect tensile strength of stone, which has been extensively applied in rock engineering and research fields for more than 30 years, is erroneous when the disc has a significant thickness [4, 17–19].

Another simple procedure to obtain the indirect tensile strength is via flexural strength tests, based on the formula for linearly elastic bodies [14]. It's known that the calculated maximum flexural stress is greater than the actual stress in the test specimen because, during the testing of a stone prism or slab under flexure, a number of factors operate to change the stress distribution in the specimen so as to reduce the maximum stress. It follows, that the tensile strength of a stone in the conditions of a flexure test is higher than in direct testing under tension [15, 20, 21].

The fact that flexural strength is physically greater is that during flexure, the maximum tensile stress is experienced only by a filament on the convex surface of the specimen, whereas during axial tension all points of the cross section experience the maximum tensile stress.

Based on the statistical theory, Weibull [14] derived relations between the strength of a brittle material under flexure, $\sigma_t$, and the axial tension, $\sigma_f$, if data from two specimens of different volumes are known:

For pure and simple flexure the following relations can be applied correspondingly:

$$\sigma_{Rt} = \frac{1}{2}\left(\sigma_{f1} + \sigma_{f2}\right) \cdot (2\mathrm{m} + 2)^{\left(-\frac{1}{\mathrm{m}}\right)} \tag{2.9}$$

$$\sigma_{Rt} = \frac{1}{2}\left(\sigma_{f1} + \sigma_{f2}\right) \cdot 2^{\left(-\frac{1}{\mathrm{m}}\right)} \cdot \left[(\mathrm{m} + 1)^2\right]^{\left(-\frac{1}{\mathrm{m}}\right)} \tag{2.10}$$

with, $\sigma_{f1}$, and, $\sigma_{f2}$, being the flexural strengths of specimens with volumes, $\mathrm{v}_1$, and, $\mathrm{v}_2$, respectively. The value of, m, is obtained as follows:

$$\mathrm{m} = \ln \frac{\mathrm{v}_2}{\mathrm{v}_1} : \ln \frac{\sigma_{f1}}{\sigma_{f2}} \tag{2.11}$$

#### 2.4.5.4 Modulus of Elasticity

There are two different procedures for evaluating the modulus of elasticity of dimension stone. The destructive or static methods, which are based on deformation of specimens under axial or flexure loads, and the so called dynamic methods, which consists on the measuring the fundamental resonance frequency.

In the first case, the specimens are tested under a uniform compression, as, for example, in the EN 14580 standard or subjected to a four-point test as in the ASTM C1352 standard. Controlled increasing test force and the corresponding deformation values allow for the determination of the Young's or elastic modulus.

In the latter case, the elastic modulus is determined either by measuring the fundamental resonance frequency as in the EN 14146 standards or by ultrasonic measurements to determine the velocity that a given ultrasonic wave crosses the length or width of a specimen.

Modulus of elasticity, compressive strength and density were the properties that Tassios and Mamillan (1985) [22] developed in relation between ultrasonic velocities in perpendicular directions, along the longitudinal or compressive direction, $V_L$, and along the transverse or shear direction, $V_T$, the dynamic modulus of elasticity, $E_{dyn}$, the Poisson ratio $\mu_{dyn}$ and the density, $\rho$.

The following equations are to be applied:

$$E_{dyn} = \frac{3V_L^2 - 2V_T^2}{V_L^2 - \frac{1}{3}V_T^2} \cdot V_T^2 \tag{2.12}$$

$$\mu_{dyn} = \frac{V_L^2 - 2V_T^2}{2 \times \left(V_L^2 - V_T^2\right)} \tag{2.13}$$

$$E_{dyn} = \rho \frac{(1+\mu)(1-2\mu)}{1-\mu} \cdot V_L^2 \tag{2.14}$$

The values obtained in this manner are in general higher than the values one may find for the static modulus. The relationship between them may range between 0.8 and 3.5 [23–26] yet, Christaras et al. [25] obtained a very good correlation between the static and the dynamic modulus of elasticity for different rock types.

#### 2.4.5.5 Creep

Creep is defined as the slow deformation of solids under constant load and for long periods of time. This time dependent deformation in rocks has been identified and fully studied since the beginning of twentieth century [27]. Studies performed by many researchers allow to conclude that stone experiment creep even at load levels well below their compressive strength but much above their tensile strength [28–31].

Dimension stone cladding design difficultly can be influenced by rock's creep unless in situation where horizontal tiling or covering may be subjected to a considerable load level. If creep deformation constraints the design, that will be due to an excessive deformation of cladding slabs under constant load conditions at a considerable level value [32].

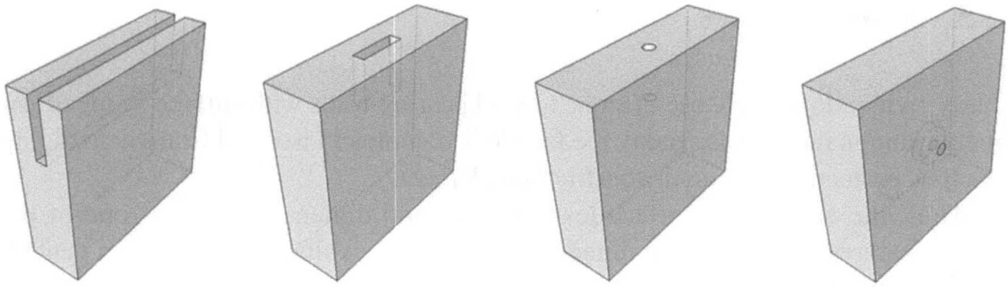

**Fig. 2.4** Dimension stone cut types; from *left* to *right*: kerf, slot, hole and undercut shapes

#### 2.4.5.6 Dimension Stone Anchorages Strength

Natural stone or other material cladding for building's façades have to allow for an inner ventilated space, the way that one of the most important aspects of designing dimension stone cladding involves determining the configuration, size and spacing of the panels and anchorages that will affix the stone panels to the underlying building substrate.

Studies that have been carried out indicate that, for the most used systems, a design focus on the bending strength of stone panels to the detriment of the anchorage strength is unsafe, yet normal practice does not take this into account [33–35].

Cuts and holes are made depending on the anchor's characteristics and type. The most common cut shapes are (Fig. 2.4) the hole for dowell (pin) insertion, a kerf cut for angles or double T anchorages; slot cuts for disks with shanks and the cylindrical hole undercut for the insertion of a cone bolt with sleeve.

The European standard EN 13364 allows for the quantification of the breaking load at the dowell hole requiring specific clamping device to hold the specimen in place and full bond between the dowell and hole filling the clearance between them. Besides being an unpractical required procedure, tests industry and designers in general claim that it's expensive and with no relevant intrinsic value. In fact, the test setup is quite hard to perform and does not reflect actual conditions in-site. Furthermore, soak specimen condition tests are not considered.

This is not the case with American standard ASTM C1354, which is a general-purpose test, simple to implement for any anchorage system and prudently recommending that the specimens should be soaked.

Both standards do not imply to address all if any safety concerns, being the responsibility on the designer to establish appropriate safety margins and to determine the applicability of the obtained results.

## 2.5 Stone Finishes

Early civilizations made use of stone to load bearing walls with surface's embellishment using manual tools. Today the façade is completely isolated from the building itself as protective skin main construction element.

Thus, the colour and surface texture of natural dimension stone is one of the major features that owners, architects and engineers explore, and people in general admire. The texture of a natural stone depends on the combination of the minerals it contains. The profiting of this aesthetical richness in stone as in any natural material depends on the manner how its surface is finished.

Natural stone may benefit from a varying degree of finishes. From smoothest to roughest, one may find different types of surface finishing: polished, honed, rubbed, abrasive, diamond sawn, flamed cleft, rustic, etc.

Depending on its hardness and natural properties, the surface of stone can be treated either by machine or using mason's tools.

A sawn surface is a raw sawn finish that has not received further treatment.

A polished surface displays a glossy or mirror finish, which brings out the full character to the stone. However, it will require extra care in order to maintain its high glow. In order to achieve the best possible effect, any holes should be filled in advance.

A honed surface has a superfine machine-ground satin finish with little or no gloss, while a rubbed surface is smooth, flat and non-reflective.

A scratched finish gives the stone an appearance of a parallel farmed field.

An abrasive finish is a flat non-reflective surface that has usually been sandblasted with silicon carbide grit, steel shot or fine glass beads.

A flamed surface is a rough surface resulting from passing a fiercely hot flame over the surface on the stone. The technique exploits the different thermal expansion properties to the particles present in the natural stone.

A cleft finish is a natural split finish in stones, which are formed in layers in the ground. When such stones are cleaved or separated along a natural seam the remaining surface is referred to as a natural cleft surface.

Bush-hammered or rustic surface is obtained with a bush hammer with a fine or coarse head. For a fine texture, the head has $7 \times 7$ pyramid-shaped teeth, for a coarse texture, $4 \times 4$ teeth.

In Table 2.5 some stone surface finishes are presented depending on the rock type. The table may not be regarded as a rule but as an indication [36, 37].

It has to be pointed out, that external facing must meet different needs other than aesthetic, and different surface textural finishes may affect the properties of cladding stone. For example a simple polish finishing on granite can reduce the strength of the stone by up to 7 % [33] and certainly affects strength increasing porosity and water absorption.

Even though some type of finishes can transform the look of the natural stone in such a way that, sometimes, the same piece of stone may appear as a completely different stone type (Fig. 2.5), the inherent reducing strength have to be taken into account in the dimension stone cladding design.

**Table 2.5** Finishes for dimension stone

| Surface texture | Granite | Marble | Limestone | Slate | Basalt | Sandstone |
|---|---|---|---|---|---|---|
| Diamond sawn | x | x | x | x | x | x |
| Polished | x | x | x | | x | |
| Honed | x | x | | x | x | |
| Scratched | x | x | x | | x | x |
| Rubbed | x | | | | | |
| Abrasive | x | x | | x | | x |
| Flamed | x | | | | | |
| Cleft | | | | x | | x |
| Bush-hammered | x | | | x | | x |

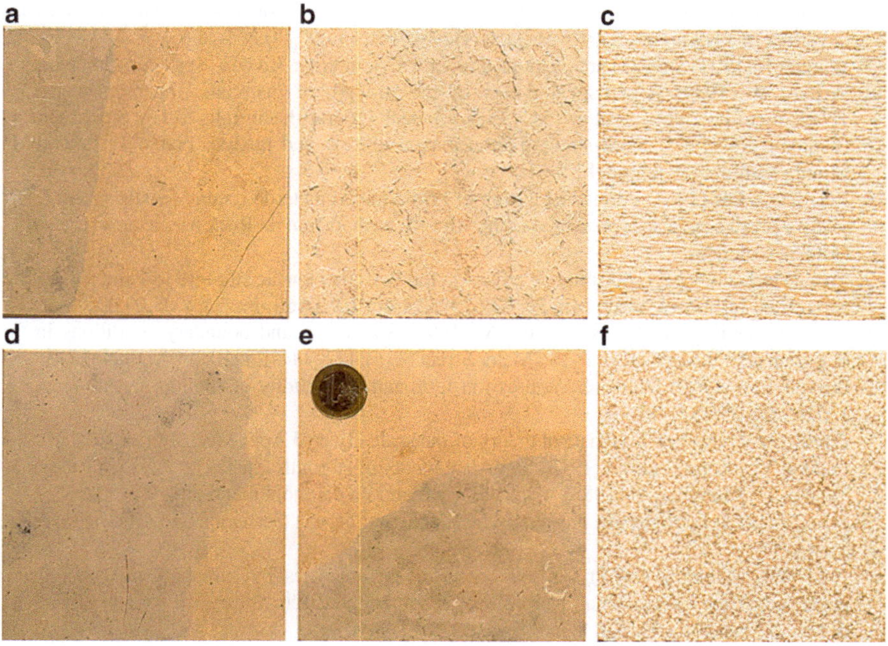


**Fig. 2.5** Effect of different stone finishes on the same limestone type; (**a**) Polished; (**b**) Flamed; (**c**) Scratched; (**d**) Honed; (**e**) Flamed and abrasive; (**f**) Bush-hammered

# References

1. Marble Institute of America (2007) Dimension stone design manual – VII. Marble Institute of America, Cleveland
2. Direcção Geral de Geologia e Energia (2006) Natural stone manual for architecture
3. Peel RF (1974) Insolation weathering: some measurements of diurnal temperature changes in exposed rocks in the Tibesti region, Central Sahara. Z Geomorph 21:19–28
4. Snethlage R, Siegesmund S, Sterflinger K et al (2010) Stone in architecture. Springer, Dordrecht/ London/New York. doi:10.1007/978-3-642-14475-2

5. Siegesmund S, Vollbrecht A, Weiss T (2003) Natural stone, weathering phenomena, conservation strategies and case studies. Geological Society, London
6. Wessman L (1997) Studies on the frost resistance of natural stone. Lund University – Lund Institute of Technology, Lund
7. López González-Mesones F (2008) El comportamiento frente al hielo de la piedra natural. Un nuevo método de evaluación Tierra y tecnología 33:54–58
8. Ingham JP (2005) Predicting the frost resistance of building stone. J Eng Geol Hydrogeol 38:387–399. doi:10.1144/1470-9236/04-068
9. Alm D, Brix S, Kjærgaard HH, Hansen KK, Grelk B (2008) Determination of resistance to salt crystallization on natural stone measured with ultrasonic wave. Salt weathering on buildings and stone sculptures. Technical University of Denmark, Lyngby
10. Smith MR (ed) (1999) Building stone, rock fill and armourstone in construction. GSL Engineering Geology Special Publications. The Geological Society – GSL, London
11. Zeisig A, Siegesmund S, Weiss T (2002) Thermal expansion and its control on the durability of marbles. Geol Soc Spec Pub 205:65–80
12. Murphy W, Winkler K, Kleinberg R (1984) Frame modulus reduction in sedimentary rocks: the effect of adsorption on grain contacts. Geophys Res Lett 1(9):805–808
13. Winkler E (1994) Stone in architecture. Properties, durability, 3rd edn. Springer, Heidelberg
14. Weibull W (1939) A statistical theory of the strength of materials. Ingeniörsvetenskapsakademiens handlingar, vol nr 151. Generalstabens litografiska anstalts förlag, Stockholm
15. Burshtein LS (1967) Rock strengths under axial tension and flexure. Fiziko-Tekhnicheskie Mining Institute Leningrad 4:53–61
16. Reichmuth DR (1967) Point load testing of brittle materials to determine tensile strength and relative brittleness. Paper presented at the 9th U.S. Symposium on Rock Mechanics (USRMS), Golden, Colorado
17. Yu Y, Yin J, Zhong Z (2006) Shape effects in the Brazilian tensile strength test and a 3D FEM correction. Int J Rock Mech Min Sci 43:623–627. doi:10.1016/j.ijrmms.2005.09.005
18. Rocco C, Guinea G, Planas J, Elices M (1999) Size effect and boundary conditions in the Brazilian test: theoretical analysis. Mater Struct 32(6):437–444. doi:10.1007/bf02482715
19. Darvell BW (1990) Uniaxial compression tests and the validity of indirect tensile strength. J Mater Sci 25:757–780
20. Davidenkov NN (1947) Statistical theory of the scale effect. J Appl Mech 14 (in Russian)
21. Frenkel YL, Kontorova MI (1943) Statistical theory of strength. Zh Tekhn Fiz 7 (in Russian)
22. Tassios TP, Mamillan M (1985) Valutazione structurale dei monumenti antichi
23. Schön J (1983) Petrophysik: Physikalische Eigenschaften von Gesteinen und Mineralen. Enke, Stuttgart
24. Judd WR, Huber C (1962) Correlation of rocks properties by statistical methods. Paper presented at the international symposium on mining research, Bureau of Mines, University of Missouri at Rolla, United States
25. Christaras B, Auger F, Mosse E (1994) Determination of the moduli of elasticity of rocks. Comparison of the ultrasonic velocity and mechanical resonance frequency methods with direct static methods. Mater Struct 27:222–228
26. Mockovciakova A, Pandula B (2003) Study of the relation between the static and dynamic moduli of rocks. Metalurgija 42(1):37–39
27. Grigs D (1939) Creep of rocks. J Geol 47(3):225–251
28. Lomnitz C (1956) Creep measurements in igneous rocks. J Geol 64(5):473–479
29. Hardy HR (1959) Time-dependent deformation and failure of geologic materials. Paper presented at the 3rd U.S. Symposium on Rock Mechanics (USRMS), Golden, CO
30. Singh DP (1975) A study of creep rocks. Int J Rock Mech Min Sci Geomech Abstr 12:271–276
31. Maranini E, Brignoli M (1999) Creep behaviour of a weak rock: experimental characterization. Int J Rock Mech Min Sci Geomech 36(1):127–138. doi:10.1016/S0148-9062(98)00171-5
32. Camposinhos RS (2009) Revestimentos em Pedra natural com Fixação Mecânica, vol 1. primeira edição edn. Edições Sílabo, Lisboa

33. Camposinhos RS, Camposinhos RPA (2009) Dimension-stone cladding design with dowell anchorage. Proc ICE – Constr Mater 162(3):95–104. doi:10.1680/coma.2009.162.3.95
34. Conroy K, Hoigard KR (2007) Stiffness considerations in dimension stone anchorage design. J ASTM Int 4:(6). doi:JAI101062
35. Camposinhos RS, Camposinhos RPA (2012) Dimension stone design – kerf anchorage in limestone and marble. Proc ICE – Constr Mater 165(3):161–175. doi:10.1680/coma.8.00052
36. Pfeifer G, Ramcke R, Achtziger J, Zilch K (2001) Masonry construction manual. Birkhäuser, Munich
37. Aalto A (1985) Architecture and urbanism. Extra edition. Gingko Pr Inc.

# Chapter 3
# Wall and Cladding Systems

**Abstract** A brief review of the presence of the stone in façades from load bearing walls to the simple skin function is presented to describe the wall and cladding methods and systems in existing and modern façades.

Adhered cladding systems to backup walls, stand alone panels and non-adhered fixing systems are described in this chapter. In all systems the main concern is to obtain an efficient water barrier pointing to the crucial function of an air space that is needed to improve the façade's performance as a rain screen.

The cladding held onto the building by an anchorage structure and a layer of insulation anchored to the grid supporting stone slabs creates a gap formed between the structure and the building thus becomes a space in which the circulation of air decreases temperature excursions, improving the building's performance.

The most used anchoring systems for natural stone cladding are briefly described.

## 3.1 Introduction

With the development of curtain wall systems, consisting of vertical and horizontal structural members, connected together and anchored to the supporting structure of the building, providing all the normal functions of an external wall, without contributing to the load bearing capacity of the building structure and the notable stone cutting technology ability, natural stone appear as thin cladding slabs or panel within lightweight curtain wall or façade systems.

Numerous techniques were employed to support the stone panels, both within curtain walls as well as individually.

Prefabrication techniques in the 1960s and 1970s brought us several solutions such as truss-type systems in which stone slabs were pre-mounted to steel trusses or stone veneer-faced precast concrete panels, both permitting faster enclosure, allowing earlier work by other trades and subsequent earlier occupancy, because each of the larger panels incorporates a number of facing pieces [1].

R. de Sousa Camposinhos, *Stone Cladding Engineering*,
DOI 10.1007/978-94-007-6848-2_3, © Springer Science+Business Media Dordrecht 2014

In the 1980s post-tensioned panels of Indiana Limestone using the same principles and standard practice of post-tensioning concrete were designed and fabricated to span 9–11 m column to column. The solid panels were constructed entirely of stone requiring no steel trusses and allowing carrying different window systems [2].

Prestressed natural stone void panels have been tested in order to span two floors in a building working in a vertical position. Being less heavy than conventional precast concrete panels, these panels may be erected using a lifting crane, thus avoiding access cradles, and are of easy fixing from inside the building [3, 4].

Façades make a major contribution to the overall aesthetic and technical performance of a building. Aesthetic on façades has a strong effect on people's judgement yet, this feature has to be regarded not only as a simple separation between the interior and exterior environment but as a complex system that is submitted to a variety of actions: physical or mechanical, such as heat, air, and moisture actions or static or dynamic loadings.

## 3.2 Cavity Wall Systems

Earlier and where the construction material is relatively porous, the wall system relies on mass to absorb and re-release the moisture. However, this may not be sufficient as any cracks in the wall will allow direct penetration. On the other hand, insulation and thermal comfort together became the key issues that have to be managed in an effective and sustainable way by buildings industry. Because the distance between particles in air and light materials, is greater than in solids cavity walls offered a solution to the buildings' envelope with some drawbacks.

Cavity walls have the advantage of providing a cavity within the depth of a wall for drainage of rainwater before it is allowed to be absorbed too far into the wall construction. Whereas loadbearing brick walls use the overall wall thickness to stop the passage of rainwater from outside to inside, cavity walls use two leaves of brickwork or other masonry units separated by a vented air gap.

The two parallel single-leaf walls are effectively tied together with wall ties or bed joint reinforcement. The space between the leaves is left as a continuous cavity or filled or partially filled with non-loadbearing thermal insulating material.

The inner leaf is generally formed with concrete block, hollow terracotta block or timber studwork. Thermal insulation is usually set on the external face of the inner leaf in order to keep the building insulated.

The same principles of cavity wall design can be applied for use when stone and concrete block or brick work is used to form the outer leaf or skin.

Natural stone can be used either as an outer leaf about 10–15 cm thick or with thinner stone bonded to the brickwork or concrete block forming a composite outer skin (Fig. 3.1).

Masonry cavity wall was developed in a need to control rain penetration through masonry walls that had become thinner. The leaves are connected with metal ties to act together in resisting and transferring the lateral loads to the building's structural

**Fig. 3.1** An external cavity wall under construction with a 15 cm thickness inner leaf on brickwork and a 12 cm thickness outer leaf on stonework

system. The cavity prevents water from reaching the inner leaf by means of capillary action and allows water to dissipate through the capillary action of the outer leaf during dry periods. When drainage holes and flashing are provided at the bottom of the cavity any water that penetrates the outer leaf is can be drained back to the outside. The holes may in some cases also act as vents, allowing vapour in the cavity to dissipate (Fig. 3.2).

Several methods or systems are available for installing stone on the exterior of buildings. Their success will depend on the manner how the environment and exposure conditions are addressed. Temperature changes, air pressure, water, in any form, as direct or indirect induced tensions have to be properly undertaken based on a solid empirical investigation encompassed with lessons from older thin-stone clad building's façade systems. In the following sections, a distinction between direct or adhered and indirect or mechanical fixing systems or methods is presented with regard to the prevailing issues about each method or system.

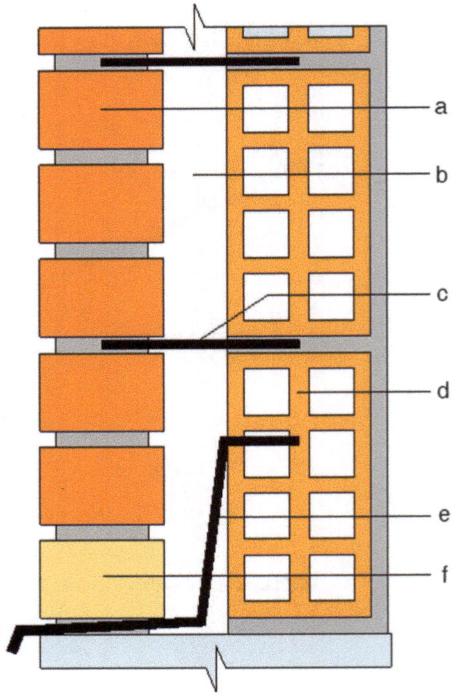


**Fig. 3.2** External cavity wall with the inner and outer leaves connected with ties; (**a**) outer leaf; (**b**) cavity; (**c**) tie; (**d**) inner leaf; (**e**) flashing; (**f**) weep hole and vent

## 3.3 Cladding Adhered Systems

Dimension stone sized as masonry units, like in Fig. 3.1, are self supporting although the labour cost to erecting the walls become expensive, the reason why composite type outer skins suit stone that is thinner.

With a narrow or well defined joint space between dimension stone slabs, cladding adhered systems rely on the chemical adhesion between the slabs and the masonry or concrete backup. This fixing method enfolds a complex system composed by the slabs, the slab's bed and the grout in the middle of the eventual filling of the joints spacing (Fig. 3.3).

This type of fixing system, using bonded to brickwork or other masonry units, assuming a continuous interlayer which completely relies in the cement chemical adhesion capability has several draw backs.

### *3.3.1 Continuous Bonding*

Although being essential to achieve complete mortar bedding at the back of the stone this is not an easy task. In fact even small voids can collect water over time, leading to premature failure. Mortar joints are also difficult to be full and

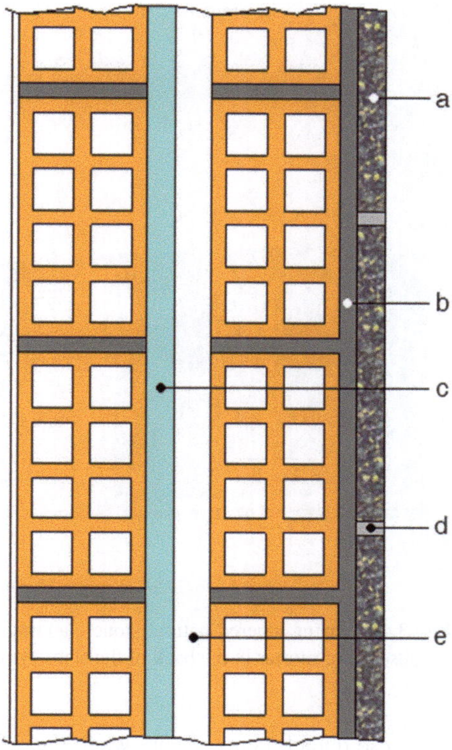


**Fig. 3.3** Cement adhered natural stone cladding in an external wall; (**a**) dimension stone; (**b**) slab's bed; (**c**) insulating layer; (**d**) slab's joint and grout; (**e**) air cavity

well-tooled, without voids or cracks that may let water into the wall. Mortar joint width should be fairly uniform since very wide joints are likely to develop shrinkage cracks and narrow joints are difficult to fill properly.

The use of occasional large stones it's in general a concern once the designer tends to obtain an appearance of load-bearing masonry as shown below for several typical stone patterns.

Cement bonding the entire back surface prevents different displacements and deformations under the same action between the slab's bed, the joint and the slab itself. This have to occur due to the fact that the physical and mechanical properties of the cladding and the grout is different, namely their coefficient of thermal expansion, hygroscopic coefficients and elastic modulus. The fact that the backup and the cladding system are rigid connected imply that the differential dimension changes are restrained giving rise to stresses in the cladding system: grout and slabs, with different distributions depending on the type and origin of the actions.

These stresses are not uniformly distributed and concentrations are to be found in the periphery of the slabs and in the joints between them. These stress concentrations are the main origin of the well-known adhesion failure of wall cladding with the inherent slab's detachment.

Failure is often instigated by small slab's cracks or slab's bowing with subsequent water and moisture penetration which with time and neglect deteriorated the bedding and joints even with an adequate mortar application and stone placement.

**Fig. 3.4** Cement adhered natural stone cladding detachments. *Lighter coloured* region indicates the absence of mortar in the back of the slabs symptomatic of poor bedding

These detachments occurs when the adhesion strength between the slab and the backup induced is not enough or able to resist to the induced stresses imposed by the restrained deformations or direct external actions (Fig. 3.4).

### 3.3.2 *Spot Bonding*

In this method an epoxy adhesive is used to bond the cladding to the backup wall. The distinction is that the epoxy is only applied to approximately 10 % of the area in four or five bonding spots. This method of stone cladding uses bi-component high strength epoxies specifically designed for this application.

The fact that the bond area is too small comparing with the total slab's area requires that the substrate has to be a compatible strength. The stress concentrations in the contacting points are extremely high for common cement based mortars or concrete strengths and detachments can occur unless the tiles dimension and thickness are small enough.

Nevertheless the concrete substrate has to be properly prepared avoiding curing compounds that may act as bond breakers and also the back of the slabs must be properly cleaned before application. The gaps of air between the stone and the substrate may reduce the potential for water staining if an appropriate installation is achieved, otherwise these voids becomes little pockets of water or moisture leading to efflorescence and other moisture-related pathologies.

**Fig. 3.5** Façade appearance after a repair intervention with visible 8 mm screw in the centre of granite slabs previously attached with a spot bonding in a poor bedding

Having clearly written specifications providing very specific quality assurance requirements to ensure that the correct products and methods will be used for the intended purpose is obviously crucial. The system may not be appropriate for cladding which dimension or weight is somehow greater than those that use smaller or thinner slabs as is the general case. The non continuous bonding in heavier dimension stones will demand higher substrate capacity strengths. If the substrate capacity is not compatible with this method drastic remedy actions are required which may be lead to unaesthetic solutions (Fig. 3.5).

## 3.4 Stone Faced Precast Panels

Large prefabricated concrete components to enfold building structures thus decreasing construction time and reducing costs and in the same time using natural dimension stone for facing the precast panels is also a system that can be used with good results. A precast concrete backup system permits faster enclosure and earlier occupancy. The size and weight of the panels may encompass limited handling facilities and as in normal precast concrete panels, they can span column to column or floor to floor.

Slender slabs can be used, but the anchoring points have to be placed closer together. The facing stone tiles have a thickness to a width ratio smaller than those used for conventionally set stone with the maximum size generally determined by the stone strength. Minimum recommended thickness is 30 mm or anchors will certainly be reflected on the exposed surface and excessive breakage or permeability problems would occur. The maximum area of the panels is about 1.4–2.8 $m^2$ [1].

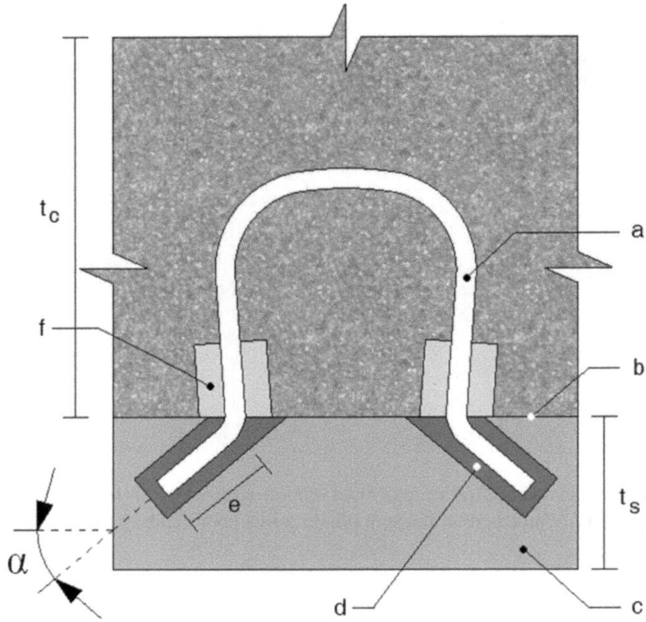


**Fig. 3.6** Cross typical section of a precast stone faced panel; Legend: $t_c$ precast concrete depth; $t_s$ stone thickness; (**a**) stainless steel anchor; (**b**) bond breaker; (**c**) stone slab; (**d**) epoxied hole; (**e**) hole depth; (**f**) compressible sleeve

The stone slabs are placed at the bottom of a casing with the finished face surface facing down. As illustrated in Fig. 3.6 on the back of the slabs pairs of holes at angles between 30 ° and 45 ° are drilled to receive clips or dowells made from stainless steel rods. Once the slabs forming the panel have been placed in the casing a bond breaker is materialized with a sheet of polyethylene or other resilient and waterproof material. The rods are inserted in the holes pre-filled with epoxy and the concrete is then cast covering the reinforcement and the clips or dowells. The connection between stone and concrete is ensured by the clips or dowells without any direct surface contact between the stone slabs and the concrete.

The anchor metal rods are sleeved with a compressible material, e.g., neoprene to accommodate differential movements allowing the anchors to flex slightly.

## 3.5 Prestressed Standalone Panels

A fundamental change in natural stone application for building façades encompassing an eco-efficiency approach where costs can be saved and environmental parameters targets may achieved by means of applying prestressing technology to natural stone.

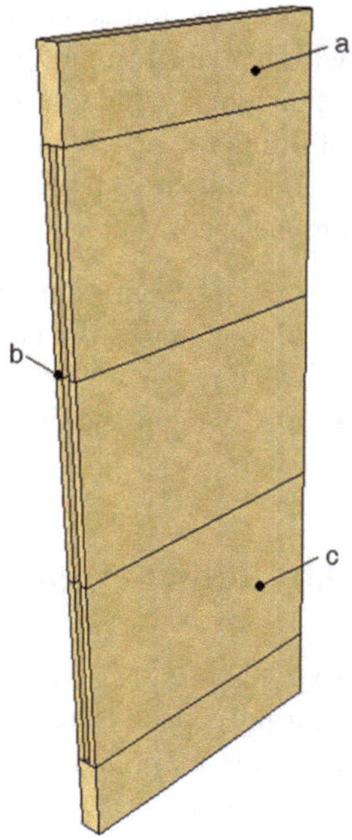


**Fig. 3.7** Three dimensional sketch view of a prestressed natural stone panel. (**a**) shut slab; (**b**) void; (**c**) inner slab

Prestressed façade panels obtained by joining together dimension stone slabs have been tested successfully [3, 5]. The slabs are connected throughout their thickness and submitted to prestressing forces with internal tendons to obtain a look like monolithic element. Two types of slabs are considered: The inner slabs and the outer slabs or shut slabs. As sketched in Fig. 3.7 the inner slabs are thinner and connected in pairs.

Their thickness and the void between them have the same total thickness of shut slabs at the panel ends, which is the panel's thickness.

Cylindrical holes in the shut slabs provide the clearance to the prestressed tendons to be tensioned allowing for the relative movement to the natural stone slabs.

The transfer of tension to the stone slabs is achieved by steel ties acting against steel anchors implanted in the shut slabs holes. In this way each tie can destress itself and burst out of the panel if damaged.

The total panel thickness, depending on the situations and desired performances, varies from 70 to 90 mm with the inner slabs individual thickness from 20 to 30 mm. The void clearance thus obtained has 30–50 mm.

The medium self weight of the panels per square meter is between 1.0 and 2.0 kN depending on the above mentioned slabs thickness and natural stone specific weight.

The prestressing forces can be achieved using bars, tendon or wires in order to have the stone in a compression stress state for the serviceability limit states.

The panels can be mounted in the buildings structures to realize a single leaf wall or the exterior leave of a cavity wall performing like non load bearing wall leaf.

Thermal and acoustic insulation for the panels is achieved forming a sandwich with a foam insulating layer in between. Furthermore with this the post-rupture behaviour becomes very similar to that of a laminated glass due the favourable effect of filling the panel void with polyurethane foam [5].

In comparison with stone faced precast concrete panels advantages must be emphasized:

- Less CO2 emission due to the fabrication of cement needed to the concrete.
- Maintenance cost is drastically reduced due the absence of bonded steel to the stone.
- Less weight and thus more economy in erecting and mounting.

## 3.6 Rain Screen Systems

External masonry walls being relegated to the function of simple leaves with small bearing capacity are not thick enough to act as mass "watertight" barriers. The available mass and volume is not enough to fulfil this requirement. Wet and dry cycles observed in external leaf walls shows that in the wet periods the humidity can't be impaired as in massive walls.

One way to prevent water penetration is to avoid the direct fixing of the cladding to the wall leaving a pressure equalized or normal ventilated air space between the cladding and the wall or substrate. This is normally called the ventilated system and its primary function is to protect the building from weathering and particularly from the infiltration of rainwater into the building walls acting as a rain screen (Fig. 3.8).

Spacing out the cladding elements from the wall also creates a ventilated air gap which, combined with the action of an insulating layer applied on the building walls, considerably improves the building's thermal efficiency. Further important advantages of the system include the dispersion of water vapour through the walls, the improvement of sound proofing as well as the reduction of maintenance operations.

In terms of thermal energy, ventilated façades bring a considerable reduction of the amount of heat that buildings absorb in hot weather due to partial reflection of solar radiation by the covering and the ventilated air gap, along with the installation of insulating material. This reduces the running costs associated with air conditioning, which in turns decreases the amount of greenhouse gas emissions. On the other side, in winter, ventilated façades maintain heat, resulting in lower heating costs thereby creating also an environmentally solution.

Ventilated systems are most common in homes and small commercial buildings. They should not be confused with vented systems, which are open only at the bottom; these systems promote drainage but do not have sufficient airflow to enable convective ventilation.

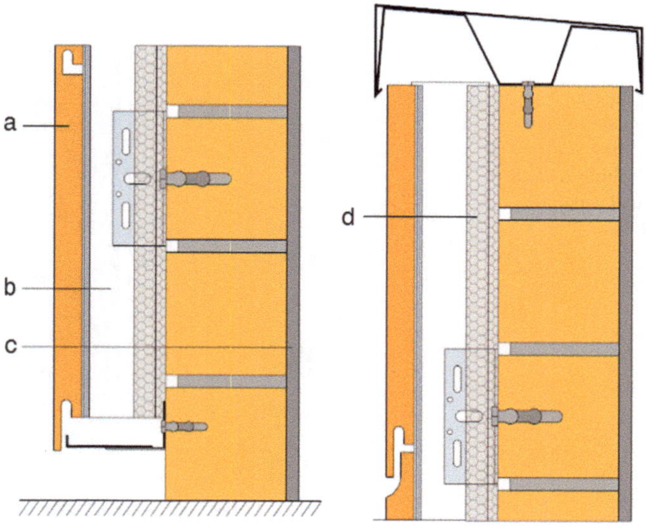


**Fig. 3.8** Ventilated wall system top and bottom cross section. (**a**) Cladding; (**b**) Air space; (**c**) backup wall; (**d**) Insulating layer

Both vented and ventilated systems assume that moisture will inevitably get behind the cladding. In contrast, pressure-equalized rain screen systems are designed to prevent moisture from entering in the first place.

Instead of relying on a continuous ventilation cavity, a pressure-equalized system is compartmentalized into discrete chambers. These ventilation cavities respond to constantly changing wind pressure. The basic configuration, incorporating two layers, or wythes, separated by an air space, comprehending variations to provide different levels of rain protection effectiveness (Fig. 3.9). A distinction should be made between the drained cavity wall, the simple or open rain screen, and the pressure-equalized rain screen wall. What is usually meant by a "rain screen wall" is generally the latter: an exterior cladding, a cavity behind the cladding, drained and vented to the outside; an inner wall plane incorporating an air barrier; and a set of compartment seals limiting the cavity size.

Pressure-equalized rain screen represent an advancement of the basic rain screen systems. As wind acts on a wall face, air passes through vents into the cavity behind the cladding. This air being enclosed appropriately by subdividing the drainage cavity with compartment seals, an equalization of pressure occurs across the cladding, thus reducing the pressure driving forces. Pressure-equalized rain screen systems are appropriate for use on all exposures and offer the highest performance potential with respect to water management.

Interface conditions between building envelope materials, components and systems should be fully detailed in a way that is both technically sound and serviceable. As in any system, detailing should, at a minimum, allow for coordination of drainage

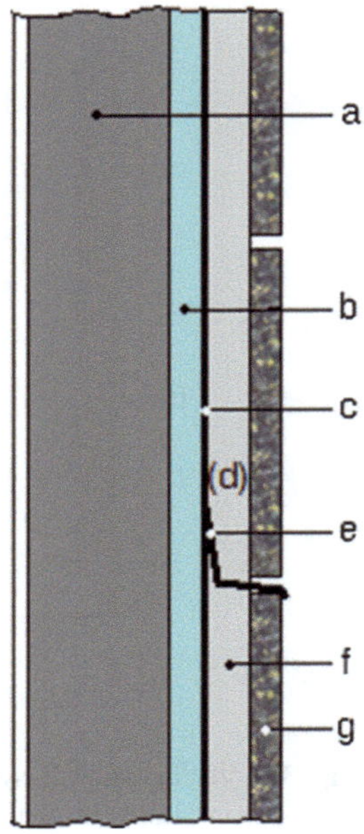


**Fig. 3.9** Pressure equalized wall system section. (**a**) backup wall; (**b**) insulating layer; (**c**) drainage plane; (**d**) air chamber providing back venting; (**e**) flashing with drip edge; (**f**) vertical separator; (**g**) cladding

planes when two or more different wall types are used in the same façade; allow for thermal and moisture-induced changes in material properties and differential thermal movement without compromise to the weather-tight integrity and thermal performance of the building envelope.

The mechanically anchoring of cladding besides being an effective fixing system gives way to an easy installation of rain screen systems either ventilated or pressure equalized.

## 3.7 Mechanical Anchored Fixing Systems

The benefits offered by mechanical anchored fixing system include reduction of the risk of cracking and detachment; easy installation; the possibility of maintenance and work on individual panels; protection of the wall structures from the action of atmospheric agents; elimination of thermal bridges and of surface condensation. The architectural design of the façade cladding can thus make use of a new generation skin capable of combining the technical performance resulting from the most advanced research in construction with the talent for composition characteristic of classic architecture.

Today there are a number of accepted conventional methods for cladding exterior building walls with natural stone.

Traditional fixing methods for the dimensional stones currently used in most European countries depend on "adhesion", through continuous superficial contact or by locally bonding between the support surface and the stone cladding. As previous mentioned incompatibility between the cladding element and the support base is behind a number of drawbacks with this type of solution.

These problems are often aggravated by the thermo-hygrometric conditions of the site and building. Safety is a crucial issue, since any detachments can result in the cladding falling with consequent risks for pedestrians and occupants of the building. Due to such considerations, most "designers" tend to avoid any direct fixing of ornamental rock cladding to façades. This is the main reason why the traditional "glued" system is being abandoned in favour of the so called "indirect" mechanical anchorage methods which are briefly described here.

As mentioned in Chap. 2 cuts and holes are made as function of the type of anchor (Fig. 2.4).

Resistance to lateral loading, which is mainly due to wind and seismic actions, is usually achieved by means of stainless steel anchors inserted into kerfs or holes drilled, or cut, into the edges of the stone panels. These anchors are connected to the building structure by mechanical means, thus providing the essential mechanical connection between the stone and the structure. One structural weak point in this type of stone construction is to be found at the kerfs or anchor holes in the edges or back of the stone slabs. Such cuts need to leave sufficient stone thickness to provide the necessary strength to resist the various winds, seismic and self-weight forces that act upon the stone panels.

Calculation of this strength is now a requirement, given that stone properties vary from stone to stone, and piece to piece. Different types of stone have different physical and mechanical characteristics.

Crucial to this process is the information gathered from a range of sources which include material strength testing, anchorage strength testing and knowledge of the stress states created within stone cladding panels by loaded anchorages.

Laboratory testing and field testing of samples of specific stones need to be performed to determine those mechanical values and properties that are to be used in structural or engineering calculations.

In the design of stone cladding, these calculations determine the correct thickness of stone to be used as a function of both the behaviour of the anchors in the edges or faces of the stone panels and that of the panel slab as a whole, providing that an accurate or realistic safety factor for the specific type of anchor and or stone used in a given project is determined.

Companies will have to submit their products to a full series of supervisory activities consisting of both an initial set of tests and production controls over their manufacturing facilities. Every stone company will need to ensure that the quality level of its finished products is equivalent to that set during the initial product tests. More specifically, these oversight procedures will involve a set of inspections of the geometrical and aesthetic features of the stone product as well as its physical and mechanical properties. The former are already subject to the company's own quality control mechanisms, and there is, therefore, nothing

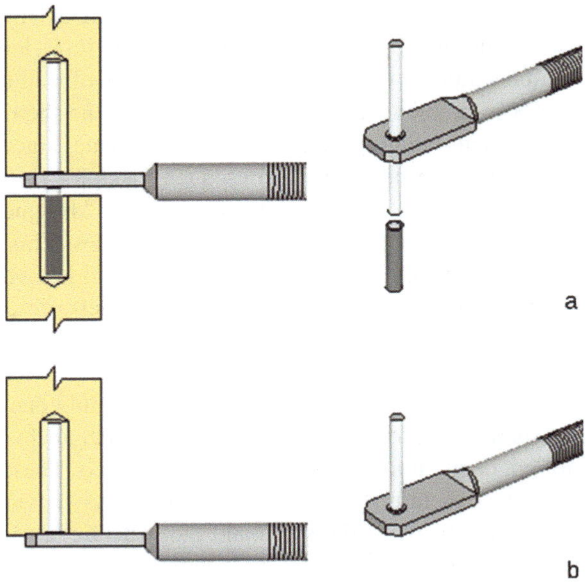


**Fig. 3.10** Pin and dowell system configuration. (**a**) Support anchor with loose pin and sliding sleeve; (**b**) support anchor with half-pin pressed

new here for stone manufacturers. The determination of physical and mechanical properties is, however, less frequent and consequently many companies will need to set up proper facilities for this, particularly at the beginning.

Theoretically, such tests could be conducted by the company's in-house test laboratory. However, given that such laboratories are only to be found in larger companies, a different scenario is certainly to be expected. Very simple tests, such as the determination of water absorption, may be conducted at the factory by means of a control system designed to test any change in the material's basic properties. More complex tests will be conducted by a laboratory specializing in the testing of stone materials under specific standards.

It should be noted that cladding exposed to the environment, particularly in large industrial urban centres, can undergo alterations as a result of atmospheric pollution caused by such agents as sulphur oxide, nitrogen oxide, carbon monoxide and carbon dioxide. As a consequence, some rocks are not suitable for external building façade cladding as their chemistry may interact with these pollutants.

That being said, the most used anchoring fixing systems are briefly described in the next sections.

## *3.7.1 Dowell Anchorage System*

In Fig. 3.10 one of the most commonly applied solutions is illustrated. It's commonly designated as the "hole and dowell" support system which is suitable for vertical surface cladding configurations for façades.

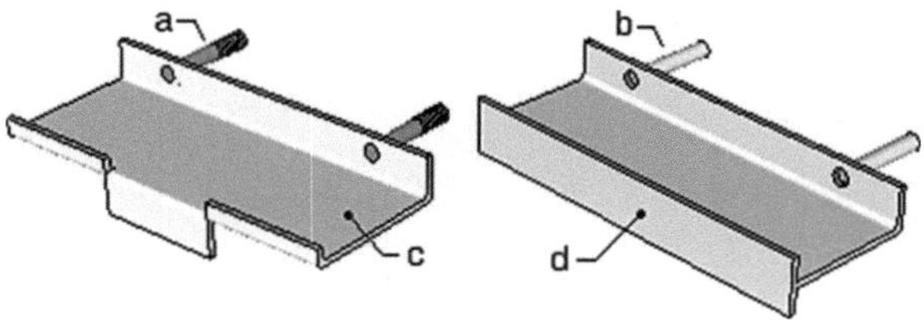


**Fig. 3.11** Kerf anchor profiles. Fixation to the backup structure with (**a**) fixation bolt or using a threaded stud (**b**); discrete length configuration in formed stainless steel (**c**) or in a continuous configuration using aluminium extruded profiles (**d**)

Special attention should be given to the number and arrangement of rods. There are always four for each panel, two per edge for edges to be mounted horizontally or vertically. Each rod engages with one or two dowells or pins, the self-weight load must be carried by the two bottom rods or pins, whether on the vertical or horizontal edges.

The lower rods need to be more resistant, as implied above, because they must bear the whole of a plate's self-weight. It is recommended that there are no more than two holes per edge so as to avoid complex overstress states resulting from the misalignment of holes [6].

### 3.7.2 Kerf Anchorage System

A kerf is a saw cut groove or slot in the edge of a stone panel. A kerf clip or kerf bar is a flat bar or thin plate configured to engage a saw cut slot in the stone edge. These profiles are fastened to a support frame or connected directly to the building structure by bolts or anchors, thus providing the essential mechanical connection between the stone and the structure.

Figure 3.11 illustrates typical kerf anchor configurations, normally in formed stainless steel or extruded aluminium profiles. Other metals may be used if properly protected against moisture and galvanic action. They may be continuous or discontinuous and are typically located in the top and bottom edges for easier access and alignment during installation.

The structural capacity of this type of anchorage depends essentially on the combined shear and flexural strength of the stone's fin or leg which are mainly dependent on (Fig. 3.12) the kerf slot width, $t_k$, the thickness of the stone fin, $t_f$, the depth of contact, D, and the length to which the kerf clip's leg is engaged, taking into account that the actual length is not necessarily the effective length of engagement [7].

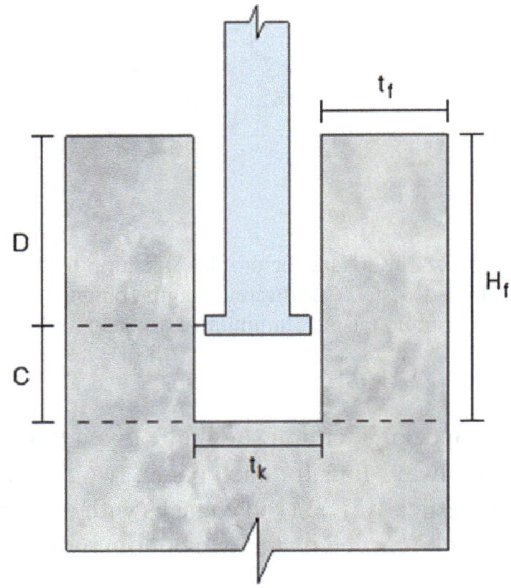


**Fig. 3.12** Kerf slot width ($t_k$), fin thickness ($t_f$) and depth of profile contact (D)

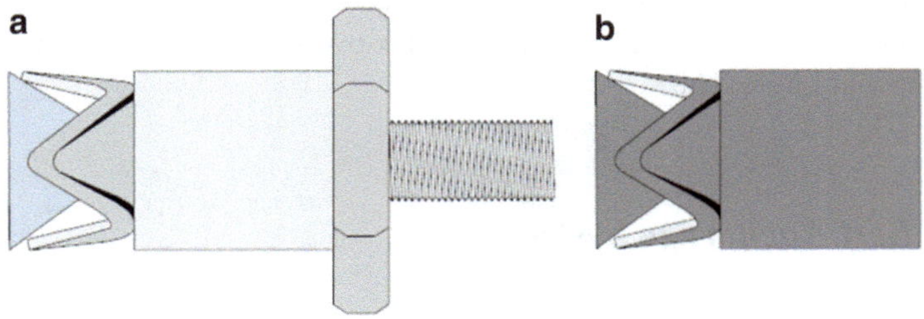


**Fig. 3.13** Fischer-type undercut anchors. (**a**) anchor with external thread and (**b**) with internal thread

## *3.7.3 Undercut Anchoring System*

Since anchoring is only a small part of the whole façade system, all other influencing factors must be given an equal amount of consideration for a successful design. However, to comprehend how undercut anchors are used, a basic understanding of the most important issues of this type of technology must be addressed.

There are mainly two types of undercut technologies to provide a keying type anchorage in the interior of the dimension stone or slab's thickness.

One system is illustrated in Fig. 3.13 in which the undercut anchors have a cone bolt, either with external thread or internal thread and generally 6 or 8 mm in diameter, an expansion ring with three or four convolutions, a sleeve and, optionally, a nut. Cone bolts and expansion rings are made of stainless steel. The sleeve is made of stainless steel or carbon. The nut is in stainless steel or aluminum.

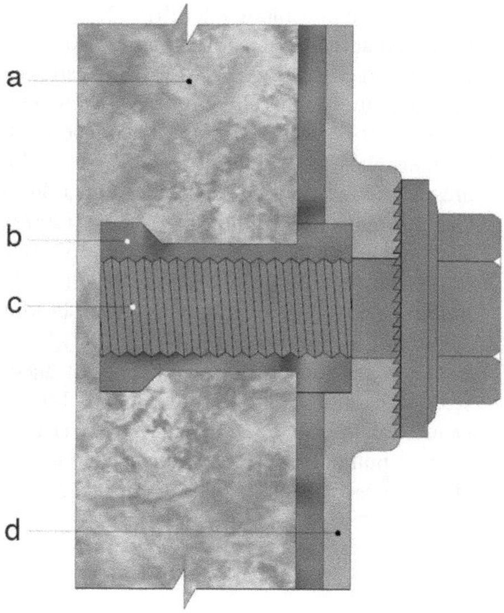


**Fig. 3.14** System using a crosswise slotted sleeve and an internal thread with a hexagon bolt. Legend: (**a**) Dimension stone; (**b**) slotted sleeve; (**c**) Hexagon bolt with internal thread and tooth lock; (**d**) panel bracket

Anchors are installed by driving the anchor sleeve against the locking ring, thus forcing it to expand within the undercut hole form and locking it within the stone, which provides a stress-free anchorage under zero applied load. This system is generally identified with the Fischer-type technology [8].

The other system consists of a special anchor made of a crosswise slotted anchor sleeve with an internal thread (Fig. 3.14).

The anchor's upper edge has a hexagon formed to it and the respective hexagon bolt with a tooth lock washer formed to it. The anchor sleeve and the hexagon bolt with a tooth lock washer formed to it are also made of stainless steel. The anchor is fitted into an undercut drill hole and, by driving the sleeve in it is deformed. The anchor sleeve is expanded to its original dimension by inserting the screw to a controlled depth, so that the sleeve sits snugly against the undercut section of the hole in the façade panel. This system is identified in general with the Keil-type technology [9].

Even though not detailed in this book, allusion have to be made to other fixing systems such as disc anchors and wire ties which have been used longer than any other technique in anchoring stone yet they are not often used in modern stone cladding systems.

## References

1. Sydney F (2000) Stone- veneer faced precast concrete panels. PCI J 45(4):72–99
2. Kluesner HF Jr (1988) Posttensioned panels of Indiana limestone. Paper presented at the new stone technology, design, and construction for exterior wall systems, ASTM STP 996, New York

3. Lello JC, Camposinhos RS (2010) Natural stone panels – an innovative façade system. Paper presented at the global stone congress 2010, Alicante, Spain
4. Camposinhos RS, Lello JC (2010) Prestressed Façade Stone Panels. In: Odasi E-TMM (ed) 7th international marble and natural stone congress of Turkey, Afyonkarahisar, Turkey, 2010. Ekim-2010 TMMOB Maden Muhendisleri Odasi, Ankara, pp 49–54
5. Camposinhos RS, Lello JC, Trigo Neves M, Santos RP (2013) Structural performance of prestressed façade limestone panels. Int J Sustain Mater Struct Syst (in print)
6. Camposinhos RS, Camposinhos RPA (2009) Dimension-stone cladding design with dowel anchorage. Proc ICE Construct Mater 162(3):95–104. doi:10.1680/coma.2009.162.3.95
7. Camposinhos RS, Camposinhos RPA (2012) Dimension stone design – kerf anchorage in limestone and marble. Proc ICE Construct Mater 165(3):161–175. doi:10.1680/coma.8.00052
8. Deutsches Institut für Bautechnik (2009) European technical approval ETA-05/0266 – Fischer-Zykon-panel anchor FZP(—W). Special anchor for the rear fixing of façade panels made of selected natural stones according to EN 1469. fischerwerke GmbH & Co. KG, Berlin
9. Deutsches Institut für Bautechnik (2004) European Technical Approval ETA-03/0055 – Keil Hinterschnittanker KH. Special anchor for the fixing of façade plates from their back side made of dust-pressed ceramic plates (stoneware) according to EN 176. KEIL Werkzeugfabrik, Berlin

# Chapter 4
# Limit States Design

**Abstract** Allowable stress design has been in use for decades for dimension stone design and it continues to enjoy popularity among engineers, even though this simplified approach appears to insufficiently cover all aspects required for a more accurate analysis. In fact, only in some cases where bending stresses prevail the calculated overall thickness at mid-span appears to be determinant.

Following a limit state approach, partial factors of safety are proposed depending on the types and on the coefficients of variation of the distributions of resistances. Their values are determined using structural reliability analysis for the load and resistance factor design format.

An application example is used to illustrate both methods, and conclusions are drawn.

## 4.1 Introduction

A skilful combination or matching of dimension stone blocks, veneer panels, tops, etc., beautifully combines natural stone's variety with man's design. In contrast to the uniformity of materials produced by machines or assembly lines, dimension stone's naturally varied appearance is a magnificent work of art and the term "uniformity of material", if applied to natural stone, has a relative meaning.

Today's thin stone veneers must be designed to resist, besides the self weight, high wind pressures and induced seismic forces.

Most of the time, the stone must accommodate hygrometric differential movement, deflection, vibration and creep of the support concrete structure, in addition to weathering and deterioration of the actual cladding stones.

Allowable stress design (ASD) has been in use for decades for dimension stone design and it continues to enjoy popularity among engineers engaged in stone cladding design. In allowable stress or "working" stress design, dimension

R. de Sousa Camposinhos, *Stone Cladding Engineering*, 
DOI 10.1007/978-94-007-6848-2_4, 

stone stresses computed under service (or working) loads are compared to some predesigned stress value expressed as a function of the flexural strength divided by a global safety factor.

This safety factor is introduced to account for the effects of overload, under-strength and approximations used in structural analysis, among other aspects. In this last case some standards and authors recommend different factors depending on whether the slab thickness is calculated for wind load considering the stresses at mid-span or for the lateral anchoring where full-scale anchoring system laboratory tested is recommended [1–3].

Over the past five decades, there has been a gradual move towards limit state design (LSD) or load/resistance factor design for the so-called "man-made" construction materials. Limit state design requires the "structure" to meet two principal criteria: the ultimate limit state and the serviceability limit state. A limit state is a set of performance criteria (e.g. vibration levels, deflection, strength, stability, buckling, twisting, collapse) that must be met when the structure is subject to loads or any type of direct or indirect action.

Limit state design has replaced the older concept of allowable stress design in most civil engineering aspects. As a result, all modern buildings are designed in accordance with a code based on the limit state theory. For example, Australia, Canada, China, Indonesia and New Zealand and almost all European countries utilize the limit state theory to develop their design codes. In the purest sense, it is now considered inappropriate to discuss safety factors when working with LSD, as there are concerns that this may lead to confusion.

Global safety factors recommended by standards, (ASTM – American Society for Testing Materials [4, 5] range from 3 to 12 and are followed by stone industry associations [3, 6]. Yet it must be said that safety factors are intended to account for: applied load variations; section size variations; material strength variations; loss of strength through time; workmanship errors. The use of safety factors is a given in all engineering practices; the method of determining the factor size, however, is a subject of ongoing discussion among architects and engineers.

When comparing equivalent values in modern codes, although partial safety factors for LSD are inferior to the so-called allowable stress design (ASD) factors, this does not necessarily mean that the latter result in higher safety [7–12].

## 4.2 State of Art

More than 30 years after the Commission of the European Communities (CEC) decided on an action programme in the construction sector to eliminate technical obstacles hampering trade and the harmonization of technical specifications, several codes have been introduced establishing a set of common technical rules for the design of buildings and civil engineering works which replaced the individual rules of the various member states. There are quite a few structural codes, the so-called Eurocodes, covering the design of structures such as those made of concrete, steel, composite steel and concrete, timber, masonry and aluminium. Astonishingly, natural stone has been practically overlooked by this programme, bearing in mind that current rules in nearly all, if not all, countries may be considered archaic or at least obsolete.

Recently the American National Standards Institute (ANSI) called upon the stone industry to develop standards that could be adopted into code. The intent is to protect the public avoiding failures that have flourished from the use of systems in untested solutions, replacing past reliance on judgment without adequate experience by uniform standards that include fundamental principles for the design and installation of contemporary systems.

To develop the new standard code requirements for ANSI, professionals who represent all interests of the stone building process are compiling the fundamental aspects of recommended stone cladding practice that should be mandatory.

The code contents will address minimum material properties by stone type, engineering evaluation, attachment types, safety factors, joint design, etc. [7].

When assessing civil engineering structures, the capability of a designed system to respond to project requirements or to meet user demands must be assessed though without overlooking some basic but fundamental principles.

Although limit state design does not seek to identify the overall safety factor applicable to each design case, it is often calculated by engineers for ease of comparison with the traditional allowable stress design, particularly in areas where limit state design is applied to ensure safety according to statistical concepts and, thus, to establish design rules or even a code.

Discussions on basic requirements for this purpose have been held at diverse instances [13–17]. There are principles that should be regarded in a design code or in establishing design principles. These same principles are dealt with in the next sections.

## 4.3 Risk and Reliability Analysis

A system can fail to perform its intended function for one or more reasons, such as natural hazards or lower performance than predicted. Failures may even include such rare events as the collapse of major structures. Although the assurance of a system's safety is primarily a task of engineers, the accepted levels of risk are subject to economic and social constraints and in this way social issues play an important role in the analysis of civil engineering systems, because these systems are more directly involved with the public than are other engineering systems.

To analyze a system's risk of failure, one must clearly identify the input to the system and its consequent response. In the case of a building, structural safety depends on the maximum load that may be imposed during the building's lifetime, and also on the load-carrying capacity, or strength, of the structure or its components.

This is a general problem which is also related to dimension stone cladding design with mechanical anchorage. In fact, predicting the maximum load and actual strength of stone cladding is subject to the same uncertainties; one cannot ensure its absolute safety, and engineers have to rely on some probabilistic concept indicating the likelihood that the available strength will adequately withstand the maximum load over the lifetime of the façade cladding system.

The said reliability is defined as the probabilistic assessment of the likelihood that a given system will perform adequately for a specified period of time under known operating conditions or, more simply stated, a system's reliability is defined as the probability of non-failure during a system's specified lifetime. The risk, on the other hand, is defined as the probability of failure under the same conditions.

As such, the risk of a system's inability to meet the respective demand is defined as the probability of failure, $P_f$, during the specified system's lifetime under specified operating conditions. System reliability, denoted by, $r$, in Eq. (4.1) is the complementary probability of non-failure.

$$r = 1 - P_f \tag{4.1}$$

A system's capability to perform under given requirements can be defined in the present case using the terms "strength" or "capacity" and "action" or "load". It's recommendable to use the term "action" in preference to "load" since any load is always an action and the opposite is not totally and suitably applicable.

### *4.3.1 Measures of Reliability*

As already mentioned, assessing risk and safety in cladding design is traditionally based on "allowable factors of safety"; these are estimated from previous experience regarding the behaviour of a particular fixing system or from observed behaviour of similar systems. A common safety factor measure applied by designers is the ratio between the assumed nominal values of the material strength, say resistance, $R^*$, and action, $S^*$:

$$F^* = \frac{R^*}{S^*} \tag{4.2}$$

For example, if the allowable stress in a dimension stone slab is 4.5 MPa and the design stress, due to wind load, is 3.0 MPa, the conventional safety factor is 1.5. The engineer may assume that the designed slab thickness is satisfactory if the calculated safety factor is greater than an accepted minimum value, which can be based on experience or on design-imposed prescriptions. If a safety factor of 1.5 is regarded as low, the engineer should redesign the slab through two different procedures: increasing the thickness and thus its capacity; decreasing the distance between supports and, consequently, reducing induced wind stress.

Because the nominal values of both resistance $R^*$ and action $S^*$ cannot be assessed with certainty, the capacity and demand functions, $R^*$, $S^*$ must be considered as probability functions. Hence, the safety factor, given by the ratio $F = R/S$ of two random variables, $R$; $S$, is also a random variable.

The system's inadequacy to meet the required demand, measured by the probability of failure, is associated with the portion of the safety margin distribution whereby it becomes less than zero. That is, a given structural element will be considered to have failed if its resistance or capacity, $R$, is less than the resulting stress or demand action, $S$. The structural element's failure probability can be stated as follows:

$$P_f = P(R \leq S) = P((R - S) \leq 0) \tag{4.3}$$

Defining $f_S(s)$ and $f_R(r)$ as the probability density independent of the functions of demand $S$ and capacity, $R$ the probability, $P_f$, of system failure is given by:

$$P_f = P(R \leq S) = \int_{-\infty}^{+\infty} \int_{-\infty}^{s \geq r} f_R(r) f_S(s) \, \mathrm{dr} \cdot \mathrm{ds} \tag{4.4}$$

and taking into account that for any random variable, $X$, the cumulative distribution function is given by:

$$F_x(x) = P(X) \leq x = \int_{-\infty}^{x} f_x(y) \, \mathrm{dy} \tag{4.5}$$

Provided $x \geq y$, it follows that for the common but special case when $S$ and $R$ are independent, that Eq. (4.4) can be written in the single integral form, since there is no physical meaning in defining the distributions to negative values of $x$:

$$P_f = P(R \leq S) = \int_{0}^{+\infty} F_R(x) f_S(x) \, \mathrm{dx} \tag{4.6}$$

This is also known as a "convolution integral" whose meaning may be easily explained by reference to Fig. 4.1. $F_R(x)$ is the probability that $x \geq y$ or the probability that the actual resistance $R$ of an element or member is less than a specific value, $x$, thus representing failure. By representing all values of $x$, i.e., by taking an integral over all $x$, the total failure probability is obtained.

For only some distributions of $R$ and $S$ it is possible to integrate analytically the convolution integral in Eq. (4.6). An example is when both are normal random variables with means $\mu_R$ and $\mu_S$ and correspondingly variances $\sigma_R^2$ and $\sigma_S^2$. The

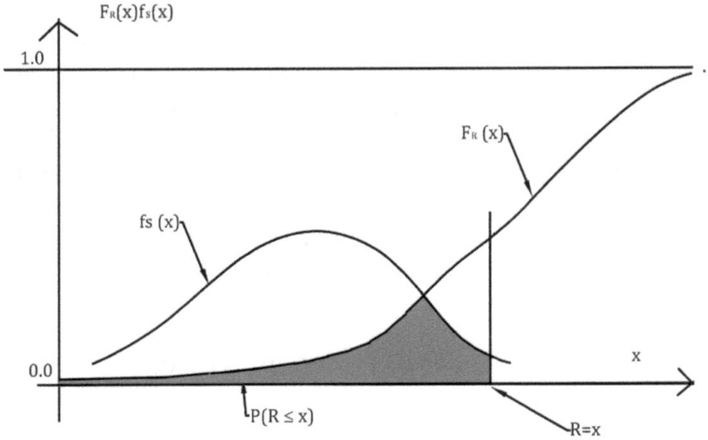


**Fig. 4.1** Basic R-S problem: $F_{R(X)}\ f_{S(X)}$ representation

safety margin given by $Z = R - S$ has a mean and variance given by well-known rules for addition (subtraction) of normal random variables:

$$\mu_Z = \mu_R - \mu_S$$
$$\sigma_Z^2 = \sigma_R^2 + \sigma_S^2 \tag{4.7}$$

Substituting Eq. (4.7) in Eq. (4.3) one has:

$$P_f = P(R - S \leq 0) = P(Z \leq 0) = \Phi\left(\frac{0 - \mu_z}{\sigma_z}\right) \tag{4.8}$$

It follows, from Eqs. (4.7) and (4.8) that this probability is equal to the probability of obtaining, in a normal distribution, values less than the mean at a distance higher than $\frac{(\mu_R - \mu_S)}{\sqrt{\sigma_R^2 + \sigma_S^2}}$, measured in standard deviations.

Considering Eq. (4.8) rewritten in the following form:

$$P_f = \Phi\left(-\frac{\mu_z}{\sigma_z}\right) = \Phi\left(-\beta\right) \tag{4.9}$$

$\beta$ can be expressed as follows:

$$\beta = -\Phi^{-1}\left(P_f\right) \tag{4.10}$$

and defined as the inverse distribution function of the standardized normal distribution of the probability of failure. Note that the negative sign on the right-hand side of Eq. (4.10) was introduced to keep $\beta$ positive for $P_f$ values of less than 0.5.

The so-called reliability index,$\beta$, is defined from Eqs. (4.8) and (4.10) based on the assumption of a normal distribution for the load effect, $S$, and resistance, $R$, and is given by:

$$\beta = \frac{\mu_z}{\sigma_z} \tag{4.11}$$

The above result is valid only under the limiting assumptions of the normal distribution of both fundamental variables, $R$ and $S$. In a more general case, when $R$ and $S$ have a general "non-normal" distribution, the probability of failure $P_f$ cannot be determined using Eq. (4.9), but only as a first approximation. In this case, however, the probability of failure $P_f$ may still be determined following Eq. (4.6) or by the use of dedicated software. Although many theoretical procedures with various degrees of complexity have been developed over the last few decades, they are not popular with practicing engineers. One reason for this unpopularity could be the lack of user-friendly software, since structural engineers without formal education in reliability-based design may not be familiar with the use of reliability based software.

Let us consider again the dimension stone slab from the previous example in which the lateral action-induced stress at mid-span has a mean value of $\mu_S = 3.0$ MPa and, for example, a variance of $\sigma_S^2 = 1.0$ MPa. The bending strength of similar slabs has been found to have a medium value of $\mu_R = 4.5$ MPa with a coefficient of variation of $V_R = 10\%$. The slab self-weight is not relevant for this effect and thus ignored.

The resistance variance is given by

$$\sigma_R^2 = [V_R \cdot (\mu_\mathrm{R})]^2 = (0.10 \times 4.5)^2 = 0.20(3) \text{ and } \mu_Z = \mu_R - \mu_S = 4.5 - 3 = 1.5.$$

The variance of the basic variable is $\sigma_Z^2 = \sigma_R^2 + \sigma_S^2 = 0.2025 + 1.0 = 1.2025$.

Therefore $\beta = \frac{\mu_z}{\sigma_z} = \frac{1.5}{1.2025} \approx 1.247401$ and, according to Eq. (4.9), the probability of failure is: $P_f = \Phi(-\beta) = \Phi(-1.247401) \approx 10.6\,\%$.

## 4.4 Design Situations for Stone Cladding

According to Eurocode EN 1990 [18], after having determined a design working life for a structure or structural component, the action variations and environmental influences during that period, the structural and material properties must be selected by taking into account distinct situations within a certain time interval implying inherent hazards or conditions. On the other hand, a differentiation of the degrees of reliability depends on the cause and mode of failure, the possible consequences of failure, the public aversion to failure and the expense and level of effort necessary to reduce the risk of failure.

**Table 4.1** Differentiation for reliability index, $\beta$, values according Eurocode 1990

| | | Ultimate limit states | | Serviceability limit states | |
|---|---|---|---|---|---|
| Consequences class | Reliability class | 1 year reference period | 50 year reference period | 1 year ref. per. | 50 year ref. per. |
| CC1 | RC1 | 4.2 | 3.3 | 2.9 | 1.5 |
| CC2 | RC2 | 4.7 | 3.8 | | |
| CC3 | RC3 | 5.2 | 4.3 | | |

**Table 4.2** Consequence class according Eurocode 1990

| Consequence class | Description | Example for building works |
|---|---|---|
| CC1 | Low consequence for loss of human life, or negligible economic, social or environmental consequences. | Agricultural buildings, sheds, greenhouses, i.e., construction or buildings where people do not normally enter. |
| CC2 | Medium consequence for loss of human life, or considerable economic, social or environmental consequences. | Construction for which the consequences of failure are significant, e.g., apartment's or office buildings, hotels, schools, access bridges, etc. |
| CC3 | High consequence for loss of human life, or very important economic, social or environmental consequences. | Construction for which the failure consequences are severe, stadiums, grandstands, theatres, high-rise buildings, bridges, dams, etc. |

It's obvious that an element which would be likely to collapse suddenly and without warning should be designed for a higher degree of reliability than one which collapse is preceded by some kind of warning thus permitting to take some measures to limit or avoid consequences. Similarly, there are issues related to the consequences of failure such as the risk to life, injury, potential economic losses, etc., that lead to a different degree of reliability.

In Eurocode 1990, three reliability classes – RC1, RC2 and RC3 – are associated with, also, three consequence classes. A link is established between these classes and values for the reliability index, $\beta$, concerning ultimate limit and serviceability limit states for 1-year and 50-year reference periods. Table 4.1 illustrates consequences and reliability classes and values for the reliability index.

As it was shown by means of Eq. (4.9) the probability of failure is directly related with the reliability.

Consequence classes are defined from low consequences to high ones depending on the end result for loss of human life, or economic, social or environmental consequences. Examples of building works that can be related with these three consequence classes are depicted in Table 4.2.

### 4.4.1 Load and Resistance Factor Design Format

The traditional deterministic measures of limit state "violation", namely the factor of safety and the load factor, can be related directly to the probability $P_f$ of limit state violation.

Generally, some upper range value of applied load or stress is compared with some lower range value of material strength. Such values might be termed "characteristic" values, reflecting that in conventional usage (e.g., in design) the load or strength is described only by this value. Thus, for instance, the characteristic tensile strength of dimension stone slabs is the strength that most (say 95 %) slabs will exceed. There is a finite, yet small, probability that some dimension stone slabs, for instance, will have an inferior strength.

For resistance, the so-called characteristic values, $R_k$, are defined on the low side of the mean resistance:

$$R_k = \mu_R \cdot (1 - k_R \cdot V_R) \tag{4.12}$$

where $R_k$ is the characteristic resistance, $\mu_R$ the mean resistance, $V_R$ the coefficient of variation for $R$ a $k_R$ constant. This is based on the normal distribution in which $R_k$ is the value of resistance below which only, say, 5 % of slab samples will fail.

Similarly, the characteristic value for the load effect is estimated on the upper side of the mean:

$$S_k = \mu_S \cdot (1 + k_S \cdot V_S) \tag{4.13}$$

where $S_k$ is the characteristic load effect, $\mu_S$ the mean load effect, $V_S$ and the coefficient of variation for $S$ and $k_S$ is a constant.

If design values are defined, for example, wind velocities, as not being exceeded 98 % of the time, a load effect is applied, then $k_S \approx 2.0537$ if $S$ is normally distributed.

In codified design, the percentiles used (such as 5 and 98 % above) are either explicitly specified or may be deduced from the characteristic value specified in existing codes or documents. Other percentile characteristic values can be obtained in the manner indicated above for normal distributions, and also for non-normal distributions.

The characteristic factor, $\gamma_k$, is defined as the lower 5 % fractile of the ratio of $F_R(x)$ and the upper 5 % fractile of $F_S(x)$. The design factor of safety, $\gamma_d$, may be defined as the ratio between the lower 5 ‰ fractile of $F_R(x)$ and the upper 5 % fractile of $F_S(x)$.

Through expression (4.6) it is possible to relate the safety factors with the probabilities of failure. This relation depends on the types and on the coefficients of variation of the distributions of resistances and load effects. For determining these relations, several types of distributions may be considered. This is the basis in structural reliability analysis of the Load and Resistance Factor Design Format (LRFD).

The design procedure using the limit state concept consists of setting up structural and load models for relevant ultimate and serviceability limit states which are taken into account in the various design situations and load cases in order to verify that no limit is exceeded when relevant design values for loads or actions and for material or product properties and for geometrical properties are used in appropriate structural and load models.

In this context the following design situations may be regarded as relevant for dimension stone design:

- Persistent situations generally related to the design working life of the element in normal use, including extreme loading conditions from wind, snow, imposed loads, etc.;
- Exceptional seismic situations applicable to elements when subjected to seismic events require a design for protection against earthquake loads.

#### 4.4.1.1 Serviceability Limit State Design

The structural design criteria used for the serviceability limit state design are normally based on the limits of deflections or vibration for normal use. In reality, excessive deformation of a dimension stone slab is normally caused by bowing, and excessive vibration or noise is caused by a defective anchorage construction technique. Certain interrelationships may exist among the design criteria defined and used separately for convenience purposes. Criteria are normally defined by established practice and economical in-service performance without excessive routine maintenance or down-time.[3–5, 19–21].

The acceptable limits necessarily depend on the type, mission and arrangement of the backup structure. Furthermore, in defining such limits, it is normally found that they are less important than the observation of good practices in construction procedures. As an example, the limiting values of vertical deflections for slabs in horizontal planes are far from being critical. In fact, good judgment by architects, engineers and contractors when specifying, designing, engineering and constructing stone and other works that interface with stone is indispensable in order to combine the stone's known performance characteristics, the building's structural behaviour and knowledge of materials and construction methods with proven engineering practice.

#### 4.4.1.2 Ultimate Limit State Design

Ultimate limit state design or load/resistance factor design for dimension stone may be applied following a suitable stress analysis taking into consideration all the relevant factors involving the design life cycle of stone cladding.

In the "load" side wind and seismic forces must be assessed according actual regulation and codes such as EC1 and EC8 [22, 23].

For resistance, a set of design rules should be based on clear and scientifically well founded, consistent and coherent theories corresponding to a good representation of the structural behaviour and of the respective materials' physics. It should be simple enough to be handled by practitioners without considerable problems and lead to conservative and robust designs. However, in a number of cases and as an alternative, more detailed design rules may be offered that consume even more calculation time, but also result in a more economical design. There are some studies applying this strategy for designing dimension stone cladding using dowell, type 31, kerf and undercut anchorages [8, 9, 11, 21, 24–28].

These studies have shown that a design focused on the bending strength of stone panels, to the detriment of the anchorage zone, is an unsafe and yet common practice. All these studies conclude that designing stone cladding systems must take into account different effects in order to evaluate the effective stress in the critical region of the anchorage geometry. In some of these studies, separate stress concentration factors were proposed to account for the anchorage zone geometry and the natural stone's specific properties.

### 4.4.2 *Dimension Stone Strength Characteristic Values*

For determining structural reliability the properties of materials together with geometrical data are an important group of the basic variables. Usually the lower value of a material property or product is unfavourable, and the 5 % (lower) fractile is then considered as the characteristic value. In the case of dimension stone it can be assumed that the theoretical model for the random behaviour of its property is known, or sufficient data may be available to determine such model.

In this case basic statistical techniques for determination of the characteristic value of stone properties may be pointed: for the flexural resistance a log-normal distribution is considered appropriate to approximate the data from the tests, and the characteristic value, $R_k$, may be obtained. In fact, using normal distribution is generally accepted if the quantity of the tested sample is greater than 50 and the skewness of the sample population is small, otherwise the CEN committee recommends the adjustment of a log-normal distribution [29, 30].

### 4.4.3 *Partial Safety Factors for Dimension Stone*

Optimization methods are available to determine partial safety and load combination factors corresponding to a predefined safety level. For simple design situations, such as for a permanent and a variable load, this is an easy task when normal distribution suitably represents the basic variables.

**Table 4.3** Partial safety factors for dimension stone cladding for three classes of consequences according to EC 1990

| | Class of consequence | | |
|---|---|---|---|
| Coefficient of variation of stone properties, $V_R$ | CC1 | CC2 | CC3 |
| <0.1 | 1.50 | 1.88 | 2.40 |
| 0.1–0.2 | 2.40 | 3.48 | 5.20 |
| 0.2–0.25 | 2.90 | 4.25 | 6.60 |
| 0.25–0.30 | 3.80 | 5.90 | 9.80 |
| 0.30–0.35 | 4.80 | 8.20 | 14.60 |
| 0.35–0.40 | 6.10 | 10.60 | 19.40 |

Through structural reliability methods, the safety formats of the design codes – i.e., design equations, characteristic values, partial safety factors and load combination factors – may be chosen so as the reliability level of all structures designed according to the design codes is homogeneous and independent of the choice of material and the prevailing loading, operational and environmental conditions. This process, including the choice of the desired level of reliability or "target reliability", is commonly understood as "code calibration".

Thus, for natural stone cladding partial safety factors may be calculated for each consequent class described in Eurocode 1990 and assuming the corresponding probability of failure from Table 4.2.

Following this principle the author [10] proposed a formulation to establish partial safety factors for dimension stone considering that:

- Wind pressure and seismic forces are assumed to be approximated by Weibull distribution with a coefficient of variation equal to 40 %.
- Flexural and tensile strength stresses, a log-normal distribution is considered.

Under these assumptions, the partial coefficient of safety for natural dimension stone design depends on the coefficient of variation of tensile strength and dimension tolerances for internal and external cladding according to existing standards or specifications [3, 31, 32].

The values of the partial safety factor, $\gamma_M$, described in Table 4.3, were determined with CodeCal © algorithm according to recommendations of Faber MH et al. [33, 34], under the abovementioned considerations for different coefficients of variation $V_R$ of stone properties and the three consequence classes.

These partial safety factors were optimized for design situations, which consider a permanent load with a partial factor $\gamma_g = 1.35$ and a single variable load with a partial factor $\gamma_Q = 1.5$ and varying ratios of permanent to total load.

It is accepted that, in the majority of situations, there are low consequences for loss of human life and low or negligible economic, social and environmental consequences. However, Table 4.3 provides partial safety factors for all three levels, CC1 to CC3, since the risk needs to be evaluated and implicitly understood by the designer when selecting a reliability index.

A designer not familiar with the difficulty and unpredictability underlying the nature of stone design will find it difficult to decide on an acceptable probability of failure. The coefficient of variation, if insufficient testing was performed, can be the first step during which designers make an error in their design assumptions.

The designer must be aware that it's very unlikely that stone cladding failure will cause high consequences for loss of human life or important economic, social or environmental consequences. This is very unlikely because this class of consequences refers to the collapse of special structures and the coefficients of variation of the natural stone's properties, as well as the partial safety factors for these applications (consequence class of CC3) are far from being plausible and affordable.

It must be emphasized that general situations correspond to a consequence class of CC1 and are given a 50-year reference period as the corresponding target index. As such, the following value for the index reliability, $\beta_{(50)} = 3.3$, is applicable which corresponds to a probability of failure of $P_f = 0.5/1,000$. This value is very close to the recommendation by Malcolm J. Faddy et al. [2].

Most building departments in the United States utilize either the 2006 or 2009 editions of the International Building Code (IBC) as the source of their building code. For structural design criteria and loadings, the IBC in turn reference ASCE/SEI 7–05 Minimum Design Loads for Building and Other Structures for much of the criteria for loads [35]. For instances a wind load factor, *LF*, shall be taken with a value of 1.6.

According to the Load and Resistance Factor Design in the United States, a strength or resistance factor, $\phi_R$, is used. In the present case, with only one material being involved, it can be thought as the inverse of the material partial safety factor, $\gamma_M$, of the Eurocode suite, so that the probability of failure is the same:

$$\phi_R = \frac{LF}{\gamma_f \cdot \gamma_M} \tag{4.14}$$

Given that and for $\gamma_f = 1.5$ the characteristic value of the dimension stone strength has to be multiplied by a resistance factor, $\phi_R = \frac{1.067}{\gamma_M}$ when a load factor of 1.6 is used.

#### 4.4.3.1 Aging and Stone Resistance Decay

It must be taken into consideration that the physical properties of stone differ widely between stone groups and even within the same stone type, and thus it's not evident that all granites are more "resistant" than marbles and that the latter are more resistant than limestone. In fact, the mineral composition, textural differences, varying degrees of hardness and pore/capillary structure are the main reasons why the same stone shows the same and uniform signs of alteration [36]. These minerals can be broken down, dissolved or converted into new minerals by a variety of processes such as frost action, thermal expansion, wetting and drying, salt decay, etc.

**Table 4.4** Aging factors for dimension stone cladding

| | Exposure conditions (Expected life cycle of 30 years) | |
|---|---|---|
| Stone type | Humid environment without frost | Humid with frost and de-icing salts |
| Fine to medium grained granite | 1.00 | 0.95 |
| Oolitic limestone | 0.90 | 0.85 |
| Fine to medium grained marble | 0.85 | 0.75 |

The decay observed in porous stones applied in building façades is caused, not only by the petrophysical properties of the stone and climatic conditions, but also by the location of the materials on the façade itself and the positioning method [37].

Several studies have been carried out to obtain coefficients that take into account the resistance decay of different types of stone [36–39]. The resistance decay of each type of stone is normally addressed in the limit state design approach by multiplying the stone resistance obtained from laboratory tests by an aging factor, of less than one, depending on the expected lifetime of the façade cladding, environment exposure and type of stone [40, 41].

Based on these studies, recommended values for an aging factor, $\eta$, take into account stone resistance decay due to aging and weathering, depending on the exposure conditions and based on a 30-year expected life cycle are depicted in Table 4.4.

#### 4.4.3.2 Design Value of the Stone Strength

The design value of the stone strength may be thus expressed according to Eq. (4.15).

$$\sigma_{Rd} = \eta \cdot \frac{\sigma_{Rk}}{\gamma_M} \tag{4.15}$$

Where:

$\sigma_{Rk}$ – Characteristic value of the stone strength;
$\gamma_M$ – Partial safety factor (Table 4.3);
$\eta$ – Aging factor applied to stone resistance decay according to Table 4.4.

## 4.5 Example of Application

In the example, a simple case for calculating the minimum thickness at mid-span of a dimension slab from a fine sized grained granite is presented using both approaches, i.e., the ASD and the LSD. The slabs span 900 mm kerf anchored in two opposite edges for a service wind load of 1.9 kPa.

**Table 4.5** Flexural strength of tested stone

| | Flexural strength ($\sigma_{str}$), coefficient of variation (c.v.) and number of tests (Qty) | | |
|---|---|---|---|
| Stone | F.S. (MPa) | c.v. | Qty |
| Granite | 9.875 | 20.9 % | 51 |

The flexural strength was determined with three-point load tests carried out on several prism samples. The bending strength and the number of tests are shown in Table 4.5.

### *4.5.1 Allowable Stress Design – ASD*

The minimum slab thickness can be obtained from the allowable stress criterion after applying a global safety factor, FS, to the ultimate strength obtained from the three-point bending test as given by the following equation:

$$\sigma_{adm} = \frac{\sigma_{str}}{\mathrm{FS}} \tag{4.16}$$

$\sigma_{adm}$ – Maximum (allowable) stress acting on the slab;
$\sigma_{str}$ – Flexural bending strength from tests;

As in American standards, a factor of safety may be applied per type of stone in an attempt to account for the variability of strength (among other factors). The global safety factor, FS, is thus determined accordingly [20, 42]. For the coefficient of variation and the type of the tested stone, we have: FS = 8. The wind induced stress at the slab's mid-span cross section corresponding to the service working load is given by:

$$\sigma_{act} = \frac{3 \times w \cdot l^2}{4 \times \mathrm{t}^2} \tag{4.17}$$

$w$ – the service value of the wind pressure;
$l$ – span length;
$t$ – slab thickness.

Making Eq. (4.16) equal to Eq. (4.17) the minimum required thickness for the slab is given by:

$$t = l \cdot \sqrt{\frac{3 \times w \cdot \mathrm{FS}}{4 \times \sigma_{str}}} \tag{4.18}$$

Substituting the corresponding values Eq. (4.18) gives: $t = l \cdot \sqrt{\frac{3 \times w \cdot FS}{4 \times \sigma_{str}}} = 0.9 \times \sqrt{\frac{3 \times 1.9 \times 8}{4 \times 9875}} = 0.0306$ m, that is to say, around 31 mm.

### 4.5.2 *Limit State Design – LSD*

For ultimate limit states, the fundamental combination of the effect of actions takes the following well-known equation:

$$E_d = \sum_{i=1}^{m} \gamma_G \cdot G_k\text{``}+\text{''}\gamma_Q \cdot \left( Q_k\text{``}+\text{''}\sum_{i=1}^{n} \psi_{0j} \cdot Q_{kj} \right) \tag{4.19}$$

where $G_k$ and $Q_k$ refer to the characteristic values of the actions (permanent and variable) and $\psi_{0j}$ is the coefficient defining the combination weight value for an action's particular effect. In general, for lateral actions affecting cladding, such as wind pressure and seismic forces, the coefficients of combination $\psi_{0j}$ are zero and the only permanent load is the slab's self weight. In this case Eq. (4.19) is expressed as follows:

$$E_d = \gamma_G \cdot G_k\text{``}+\text{''}\gamma_Q \cdot Q_k \tag{4.20}$$

It must be noted that the plus sign in the above equations can only be numerical and applied if the effect of the actions are additive for the design variable under analysis.

The safety coefficients are intended to take into account uncertainties associated with the analysis model and with the actions' intensity variation. They are usually defined in the structural codes for persistent and transient design situations, assuming, for the most common cases of building structures and for unfavourable effects, the consensual values used, for example, in the Eurocode suite are 1.5 for $\gamma_Q$ and the values of 1.5 or 1.35 for $\gamma_G$ depending on the material and the applicable specification.

In case of the wind dynamic pressure, according to the Eurocode suite, the partial safety factor, $\gamma_f$, for variable actions must be assumed with a value equal to 1.5. Thus, considering the working load as the characteristic wind pressure, the design bending moment per unit of width in the slab's mid-span cross section is given by:

$$M_{Sd} = \gamma_f \frac{wl^2}{8} = 1.5 \times \frac{1.9 \times 0.9^2}{8} \approx 0.2886\ \text{kNm/m} \tag{4.21}$$

The design flexural strength of the slab is obtained by dividing the characteristic value of the flexural strength by the correspondent partial safety factor for the stone according to Table 4.3 for the relevant class of consequence.

Given that the quantity of tested specimens is reasonably adequate, the characteristic value of the flexural stress is obtained assuming a normal distribution for the tested stone values. Thus the characteristic stress value for flexural strength, $\sigma_{Rk}$, may be obtained as follows:

$$\sigma_{Rk} = \sigma_{\text{str}} \cdot (1 - 1.64 \times V_R) \tag{4.22}$$

Substituting the corresponding values, from Table 4.5, in Eq. (4.22) we have:

$$\begin{aligned}\sigma_{Rk} &= \sigma_{\text{str}} \cdot (1 - 1.64 \times V_R) = 9875 \times (1 - 1.64 \times 0.209) \\ &= 6490.245 \text{ kPa}\end{aligned}$$

The design value of flexural strength, $\sigma_{Rd}$, for the tested stone is obtained according to Eq. (4.15) as follows:

$$\sigma_{Rd} = \eta \cdot \frac{\sigma_{Rk}}{\gamma_M} = 1 \times \frac{6490.245}{2.9} = 2238.016 \text{ kPa.}$$

The design resisting bending moment, $M_{Rd}$, per unit of width, at the slab's mid-span cross section will depend on the slab's thickness, as follows:

$$M_{Rd} = \frac{1}{6} \times \sigma_{Rd}.t^2 \tag{4.23}$$

The limit state design equation $M_{Rd} \geq M_{Sd}$ being applied substituting the corresponding values from Eqs. (4.21) and (4.23) the minimum thickness for the slab is obtained: $t \geq \sqrt{\frac{0.2886 \times 6}{2238.016}} \approx 0.0278$ m, that is to say, around 28 mm.

Although it may be discussed whether the 3 mm difference in the calculated thickness may be overlooked, there's a fundamental difference in the manner how each thickness was determined.

The issue arising from the ASD approach is that the designer is looking for a "spot" in a sea of possibilities that he cannot assess to determine the level of risk associated with its "solution".

In the other calculation method using the limit state design, the solution is not unique and the designer can decide or choose different solutions (thickness) according to each situation whilst knowing the associated level of risk.

Applying a limit state approach requires defining this variability and uncertainties by frequency distribution curves, which can be achieved through simple and affordable laboratory testing. On the other hand, a more rigorous analysis must be carried out to compute induced stresses in the anchorage support areas. Although this is not a straightforward task, research work has been carried out to validate formulae through computational analysis that permits validating specific formulae for the stress analysis of different anchorage systems. Application examples of the indicated design format are presented in further chapters.

## References


1. Yates TJS, Matthews S, Chakrabarti B (1998) External cladding: how to determine the thickness of natural stone panels. Construction Research Communications Ltd by permission of Building Research Establishment Ltd, London
2. Faddy MJ, Wilson RJ, Winter GM (2003) The determination of the design strength of granite used as external cladding for buildings. Case studies in reliability and maintenance. Wiley, St. Lucia
3. Marble Institute of America (2007) Dimension stone design manual – VII. Marble Institute of America Cleveland, Cleveland
4. Materials A-ASfTa (2010) Standard guide for selection, design, and installation of dimension stone attachment systems, vol ASTM C1242 – 10. American Standard of Testing Materials, West Conshohocken
5. Materials A-ASfTa (2010) Standard guide for selection of dimension stone for exterior use, vol ASTM C1528 – 10. American Standard of Testing Materials, West Conshohocken
6. Institute IL (1998) Limestone handbook, 21st edn. Indiana Limestone Handbook, Bedford
7. Lewis MD (2007) Status on development of code requirements for exterior stone cladding. J ASTM Int 4(7):1–4. doi:10.1520/JAI100856
8. Camposinhos RS, Camposinhos RPA (2012) Dimension stone design – kerf anchorage in limestone and marble. Proc ICE Construct Mater 165(3):161–175. doi:10.1680/coma.8.00052
9. Camposinhos RS (2012) Dimension stone design – partial safety factors: a reliability based approach. Proc ICE Construct Mater 165(3):145–159. doi:10.1680/coma.10.00018
10. Camposinhos RS, Pires V, Amaral P, Rosa LG (2011) Slate flexural and anchorage strength considerations in cladding design. Constr Build Mater 25(10):3966–3971. doi:10.1016/j.conbuildmat.2011.04.029
11. Camposinhos RS, Camposinhos RPA (2009) Dimension-stone cladding design with dowel anchorage. Proc ICE – Constr Mater 162(3):95–104. doi:10.1680/coma.2009.162.3.95
12. Beall C (1990) Selecting a safety factor for stone. Aberdeen Group, Boston
13. Virlogeux M, Walraven JC (1999) The development of an international codification for structural concrete with the CEB-FIP Model Codes. In: Colloquium I (ed) Concrete Model Code for Asia, Phuket
14. Walraven JC (2004) Thinking about codes. Struct Concr J FIB 3:93–100
15. Ditlevesen O, Madsen HO (1996) Structural reliability methods. Wiley, Chichester/New York
16. Ellingwood B, Kanda J, Council on Tall Buildings and Urban Habitat. Committee 9A/10, Structural Engineering Institute (2005) Structural safety and its quality assurance. American Society of Civil Engineers, Reston
17. Melchers RE (1999) Structural reliability analysis and prediction, 2nd edn. Wiley, Chichester
18. British Standards Institute (2006) Eurocode basis of structural design BS EN 1990:2002 includes amendment A1:2005. British Standards Institute, London
19. British Standards Institute (2010) Code of practice for the design and installation of natural stone cladding and lining. Traditional handset external cladding, vol BS 8298–2:2010. BSI, London
20. Gere AS (1988) Design considerations for using stone venner on high-rise buildings. B. Donaldson edn. ASTM – American Society for Testing Materials, Philadelphia
21. Lewis MD (1995) Modern stone cladding: design and installation of exterior dimension stone systems, vol MNL 21, ASTM manual series. ASTM, Philadelphia
22. CEN – European Committee for Standardization (2004) Eurocode 8: design of structures for earthquake resistance – part 1: general rules, seismic actions and rules for buildings vol EN 1998–1. CEN, Brussels
23. CEN – European Committee for Standardization (2002) Eurocode 1: actions on structures — Part 1–4: general actions – wind actions. CEN TC 250
24. Camposinhos RS (2009) Revestimentos em Pedra natural com Fixação Mecânica, vol 1. primeira edição edn. Edições Sílabo, Lisboa

25. Camposinhos RS, Amaral P (2007) Caderno técnico de aplicação, uso e manutenção de rochas ornamentais. Assimagra, Lisboa
26. Conroy K, Hoigard KR (2007) Stiffness considerations in dimension stone anchorage design. J ASTM Int 4(6):1–12. doi:JAI101062
27. Lammert BT, Hoigard KR (2007) Material strength considerations in dimension stone anchorage design. J ASTM Int 4(6):1–19. doi:JAI101062
28. Scavuzzo R, Acri J (2000) In-place load testing of a stone cladding anchorage system. Dimension stone cladding: design, construction, evaluation, and repair. American Society for Testing and Materials, West Conshohocken
29. CEN – European Committee for Standardization (1999) EN 12372 Determination of Flexural Strength under Concentrated Load. CEN – European Committee for Standardization, Brussels
30. CEN – European Committee for Standardization (2008) EN 13161 – Natural stone test methods. Determination of flexural strength under constant moment Natural stone. Test methods. CEN – European Committee for Standardization, Brussels
31. CEN – European Committee for Standardization (2003) EN 13373:2003 Natural stone test methods – determination of geometric characteristics on units. Natural stone. Test methods. CEN – TC 246, Brussels
32. Association NBGQ (2011) Stone granite specs. Specifications for architectural granite. The National Building Granit Quarries Association, Bare
33. Faber MH, Sorensen JD (2003) Reliability Based Code Calibration – The JCSS Approach. In: 9th International conference on applications of statistics and probability, San Francisco,
34. Vrouwenvelder T (2003) Reliability based code calibration – the use of the JCSS model code. Joint Committee of Structural Safety
35. American Society of Civil Engineers, ebrary Inc. (2010) Minimum design loads for buildings and other structures ASCE standard
36. Miglio BF, Richardson DM, Yates TS, West D (1999) Assessment of the durability of porous limestone: specification and interpretation of test data in UK practice – stone weathering and durability. Paper presented at the dimension stone design cladding: design, construction, and repair, Willowbrook
37. Bernabéu A, Cura MAGd. The influence of design requirements on the durability of porous building stones used in façades. A case study. In: Gomez-Heras DM (ed) Workshop on limestone decay and conservation, Malta, 2007. Stone Weathering and Pollution NETwork (SWAPNET)
38. Grelk B, Christiansen C, Schouenborg B, Malaga K (2007) Durability of marble cladding — a comprehensive literature review. Dimension stone use in building construction
39. Prikryl R, Smith BJ (2007) Building stone decay: from diagnosis to conservation. Special Publications, London
40. Lemvig J, Rasmussen J, Jensen JF, Hansen KK, Grelk B Natural stone panels for building façades – loss of strength caused by temperature cycles. In: Building Physics 2002 – 6th Nordic symposium, Trondheim, 2002
41. Hansen KK, Lekso H, Grelk B (2003) Assessment of the durability of marble cladding by laboratory exposure compared to natural exposure. Paper presented at the research in building physics
42. Miglio B, Redding J et al (1997) Guide to selection & testing of stone panels for external use. University of Bath/CWCT – Centre for Window & Cladding Technology, Bath

# Chapter 5
# Actions and Stone Strength

**Abstract** The value of the actions and their combination rules values are presented in order to determine their effects on building's façade and cladding systems.

Expressions are given to determine the self-weight of dimensioning stones considering the water absorption and open porosity of the corresponding natural stone slabs.

Formula and calculation examples are given to compute the external and internal pressures due to the wind action on pressure equalized, ventilated or joint sealed vented rainscreen claddings systems.

Simple formulas with application examples are presented to compute the maximum response of cladding elements attached to buildings façades when subjected to earthquake ground motions transmitted throughout the building structure.

Guidance is also given for the calculation of the joint's spacing and width regarding movements originated by thermo-hygrometric actions.

Information and guidance is also given to obtain the characteristic values of dimension stone strength based on the information that is mandatory to collect in some of the relevant standard tests.

An application example of design assisted by testing is presented.

## 5.1 Introduction

Several assumptions are necessary for predicting the structural behaviour of a given structural element, once they are simply an idealization of the true system.

Among the factors that affect the choice of a structural model some are related with the geometric properties of the elements, such as spans, supports configuration, cross-section dimensions, etc., and others are those that significantly depend on materials properties, e.g., strength, constitutive relations, temperature and moisture dependence, etc.

R. de Sousa Camposinhos, *Stone Cladding Engineering*,
DOI 10.1007/978-94-007-6848-2_5, © Springer Science+Business Media Dordrecht 2014

The third factor is related with the actions quantification and definition. Indirect or direct, static or dynamic, actions are considered separately from the structural resistance of the model.

Following this approach, methodologies and data are present to obtain the characteristic values for the relevant action in buildings' façades and to calculate the characteristic values of the relevant strength of dimension stone, considering the relevant Eurocode standards assuming the well-known rules of the safety and working requirements in a limit state theory.

## 5.2 Combination of Actions

The design values of the actions have to be derived from their characteristic or other representative values.

The design value $A_d$ of an action $A$ can be expressed in general terms as:

$$A_d = \gamma_f A_{\mathrm{rep}} \tag{5.1}$$

where $A_{\mathrm{rep}}$ represents either the characteristic value of the action $A_k$ or the combination value $\psi_{\mathrm{n}} A_k$ where the combination factor $\psi_{\mathrm{n}}$ is equal or less to 1 according to the following Equation:

$$A_{\mathrm{rep}} = \psi_{\mathrm{k}} \cdot A_k \tag{5.2}$$

The partial factor, $\gamma_f$, takes account of the possibilities of unfavourable deviations of the action values from the representative values [1].

In the following sections combination and characteristic values of the most relevant actions on façade cladding are presented allowing for the determination of their effects either at the ultimate or serviceability limit states when appropriate.

Considering its variation in time and space, self-weight of a construction element is classified as permanent fixed action while imposed load as variable free action.

Generally the imposed load is considered as static load, which may be increased by a dynamic magnification factor. If an imposed load causes significant acceleration of the structure or structural element, dynamic analysis should be applied in accordance.

### *5.2.1 Ultimate Limit States*

Limit state design is often dependent (or oriented) on a certain ultimate limit state and the format of Eq. 3.21 is simplified accordingly:

**Table 5.1** Partial factors for actions – Ultimate limit state in persistent situations

| Permanent actions | Variable actions | |
|---|---|---|
| Unfavourable effect | Favourable effect | Unfavourable effect |
| $\gamma_G = 1.35$ | $\gamma_G = 1.0$ | $\gamma_Q = 1.5$ |

$$\gamma_G \cdot G_k + \gamma_{Q,1} \cdot Q_{k,1} + \gamma_Q \cdot \left[ \sum_{i>1} \psi_{0,i} Q_{k,i} \right] \tag{5.3}$$

where $G_k$ and $Q_k$ refer to the characteristic values of the actions (permanent and variable) and $\psi_{0i}$ is the coefficient defining the combination weight value for an action's particular effect. In general, for lateral actions affecting cladding, such as wind pressure and seismic forces, the coefficients of combination $\psi_{0i}$ are zero and the only permanent load is the slab's self weight.

### 5.2.2 *Serviceability Limit States*

Limit state design regarding serviceability limit state depends mainly on the type and arrangement of the backup and on the observation of good practices in construction procedures. Criteria are normally defined by established practice and economical in-service performance without excessive routine maintenance or downtime [2–8].

Nevertheless in some situations it could be necessary to compute for example minimum joints spacing or excessive deflection in floors or ceilings. In these situations the quasi-permanent combination of action should be used in calculations. Taking into consideration that it is not likely to expected more than one variable action acting the following expression shall be is used:

$$G_k + \psi_2 \cdot Q_{k,1} \tag{5.4}$$

The coefficient $\psi_2$ is defining the combination weight value for quasi-permanent combinations.

The partial and combination coefficients are depicted in Tables 5.1 and 5.2.

Buildings' façades and its cladding are in general built and assembled along vertical planes; as such the self weight of its elements rarely mobilizes relevant out of plane flexural stresses. This is the reason why, when analysing the state of stress or deformation in vertical positioned cladding elements, only a variable action is considered, normally wind or seismic induced action. For claddings in horizontal or very inclined surfaces self weight and one or two variable action have to be combined according to expressions (5.3) and (5.4), e.g., wind and snow.

**Table 5.2** Combination coefficients

| Combination coefficients – in buildings | $\psi_0$ | $\psi_1$ | $\psi_2$ |
|---|---|---|---|
| Imposed loads on residential areas | 0.7 | 0.5 | 0.3 |
| Imposed loads on office areas | 0.7 | 0.7 | 0.3 |
| Imposed loads on shopping areas | 0.7 | 0.7 | 0.6 |
| Imposed loads on storage areas | 0.7 | 0.7 | 0.6 |
| Wind | 0.6 | 0.2 | 0.0 |
| Snow in horizontal surfaces (altitude > 1,000 m) | 0.7 | 0.5 | 0.2 |
| Snow in horizontal surfaces (altitude < 1,000 m) | 0.5 | 0.2 | 0.0 |
| Temperature (fire effect not considered) | 0.6 | 0.5 | 0.0 |

## 5.3 Characteristic Value of Actions

Further to the self-weight, the relevant forces acting on cladding are due to the wind and the seismic action. Inertial forces induced by the response of buildings to seismic events are seldom considered as well as the special emphasis is given in its effects calculation.

In addition, relevant data for the calculation of the effects of the thermal and hygrometric action and snow load is also given.

### 5.3.1 *Self-Weight*

The self-weight of dimension stone elements shall be determined considering nominal dimensions and nominal values of the rock densities provided by relevant standards and tests.

The self-weight of the dimension stone depends on the density of the rock. The term "density" is used for weight per unit volume, area or length. For materials having all three dimensions of the same order of magnitude, the characteristic values of densities are given as weights per unit volume, usually expressed in $kN/m^3$.

In general density is a random variable, which may have in some cases (e.g. in case when moisture content and degree of consolidation may affect the density) a considerable scatter. In such cases the mean value and variance may be determined using available experimental measurements. The characteristic value of the density is usually defined as the mean. However, when the coefficient of variation is greater than 5 %, then upper and lower characteristic value may be used [9].

For cladding materials such as dimension stone claddings having one dimension of smaller order of magnitude than the other two dimensions the characteristic values are expressed in weight per unit area.

In fact, for some natural rock types, the variation of the self-weight due to alteration in the humidity content depends directly on the open porosity. The effective or open porosity does not comprise all pore spaces in a sample but only voids where fluids and air can access and depends on the apparent density mass and on the weight of the absorbed water.

We can consider that open porosity values greater than 0.5 % may affect perceptibly dimension stones self-weight, particularly in unprotected rainy areas or in high humidity environment [10]. In these circumstances the self-weight shall be computed taking into consideration the water absorption coefficient at atmospheric pressure or based on the open porosity. The saturated specific weight, $\gamma_{\mathrm{sat}}$, may be found as follows:

$$\gamma_{\mathrm{sat}} = \gamma_{\mathrm{dry}} \left(1 + \frac{W_a}{100}\right) \tag{5.5}$$

$$\gamma_{\mathrm{sat}} = \gamma_{\mathrm{dry}} + \frac{p_0}{100} \cdot \left(\gamma_{\mathrm{w}} - \gamma_{\mathrm{dry}}\right) \tag{5.6}$$

where:

$\gamma_{\mathrm{dry}}$ is the dry specific weight;
$\gamma_{\mathrm{w}}$ is the water specific weight;
$p_0$ is the effective or open porosity according Eq. (1.4);
$W_a$ is the water absorption at atmospheric pressure according Eq. (1.5).

### 5.3.2 *Wind Action*

Codes and standards habitually consider a basic or reference wind speed for various locations in a given region or country. Almost always a reference height of 10 m in the open terrain is chosen along with modification factors for the effects of height and terrain type and pressure or force coefficients for structures of various shapes are also provided [11–13].

Several comparisons between major wind loading codes and standards have been made. Several comparative studies [14, 15] have been performed including standards such as the ISO 4354, Eurocode 1 part 1–4, the ASCE Standard ASCE 7-98, AIJ Recommendations for Loads on Buildings and the Australian Code AS/NZS 1170.2:2002. There are conceptual relevant differences when comparing with the Eurocode edition on wind loads [16]. Nevertheless, because it's regarded to be a representative multinational wind loading standard, comprehending several years of work by committees from many countries of the European Union and also is mandatory throughout several European countries the part 1–4 of the Eurocode 1 [17] will be followed in this section.

One of the main factors in the determination of wind actions on structures is the characteristic peak velocity pressure or the peak pressure, $q_p$. This parameter is the characteristic pressure due to the wind velocity of a given undisturbed wind field and accounts for the mean wind velocity and a turbulence component. The characteristic peak velocity pressure, $q_p$, is influenced by the regional wind climate; local factors such as the terrain roughness, the orography factor and the height above ground.

The wind climate for different regions in Europe, as in many other parts of the world, is described by values related to the characteristic 10 min mean wind velocity at 10 m above ground of a terrain with low vegetation. These characteristic values correspond to annual probabilities of exceedance of 2 % which corresponds to a return period of 50 years, the so-called basic wind velocity, $v_b$.

The values of the basic wind velocities are given for the different European countries in National Annexes. Some countries define areas for an explicit value of the basic wind velocity and other countries provide climatic maps with its values as well as the procedure to interpolate between the provided velocity isoline values.

The basic velocity and the basic pressure are related by the following relationship derived from the principles of the energy conservation:

$$q_b = \frac{1}{2}\rho \cdot v_b^2 \tag{5.7}$$

where $\rho$ is the density of air at sea level usually taken as 1.25 kg/m$^3$. This expression represents the mean velocity pressure (averaging interval 10 min.), without the influence of the turbulence of the wind and at a reference height of 10 m in open terrain with a return period of 50 years.

For that reason the basic value of the velocity pressure has to be transformed into the value at the reference height of the considered element or structure.

The velocity at a relevant height and the gustiness of the wind depends on the terrain roughness. A roughness factor describing the variation of the speed with height has to be determined in order to obtain the mean wind speed at the relevant height:

$$v_m(z) = c_r(z) \cdot c_0(z) \cdot v_b \tag{5.8}$$

where $v_m(z)$ is the mean velocity, $c_r(z)$ the terrain roughness factor and $c_0(z)$ the orography factor usually taken as 1.0 except in the cases that the buildings are located on elevations like hills, mounts or other relevant terrain prominences.

The terrain roughness $c_r(z)$ depends on a terrain factor, $k_r$, a roughness length, $z_0$ and a minimum height, $z_{min}$:

$$c_r(z) = k_r \cdot \ln\left(\frac{z}{z_0}\right) \text{ for } z_{max} \geq z \geq z_{min} \tag{5.9}$$

with:

$$k_r = 0.19 \times \left(\frac{z_0}{0.05}\right)^{0.07} \tag{5.10}$$

**Table 5.3** Terrain categories and terrain parameters

| Terrain category | Characteristics of the terrain | $z_0$ [m] | $z_{min}$ [m] |
|---|---|---|---|
| 0 | Sea or coastal area exposed to open sea | 0.003 | 1 |
| I | Lakes or flat and horizontal area with negligible vegetation and without obstacles | 0.01 | 1 |
| II | Area with low vegetation such as grass and isolated obstacles (trees, buildings) with separations of at least 20 obstacle heights | 0.05 | 2 |
| III | Area with regular cover of vegetation or buildings or with isolated obstacles with separations of maximum 20 obstacle heights (such villages, suburban terrain, permanent forest) | 0.3 | 5 |
| IV | Area in which at least 15 % of the surface is covered with buildings and their average height exceeds 15 m | 1 | 10 |

The values of the roughness length, $z_0$ and minimum height, $z_{min}$, are depicted in Table 5.3. In Eq. (5.10) the value 0.05 m corresponds to roughness length of category II terrain, ($z_{0,II}$). The value of $z_{max}$ is to be taken as 200 m.

To determine the peak velocity pressure, $q_p(z)$, at height z, which includes mean and short-term velocity fluctuations, the general formulae is presented in part 1–4 of Eurocode 1 [16]. The peak velocity depends on a turbulence factor, $k_I$, and on the orography factor, $c_0(z)$ and may be expressed as a function of the basic pressure and a exposure factor, $c_e(z)$, as follows:

$$q_p = c_e(z) \times q_b \tag{5.11}$$

In general conditions, for flat terrain, $c_0(z) = 1$ and also the turbulence factor $k_I = 1$ so that the exposure factor $c_e(z)$ may be expressed as follows:

$$c_e(z) = (c_r(z))^2 \left[ 1 + \frac{7}{\ln(z/z_0)} \right] \tag{5.12}$$

In these conditions the exposure factor $c_e(z)$ is illustrated in Fig. 5.1 as a function of height above terrain for the different terrain categories defined in Table 5.3.

The wind pressure acting on the cladding surfaces, $w_e$, is obtained from Expression (5.13) for a given reference height, $z_e$:

$$w_e = q_p(z_e) \cdot c_{p_e} \tag{5.13}$$

Depending on the geometrical aspect ratio of the façade, the reference height, $z_e$, shall be considered to be the height of the façade, either the height of the façade or the width, and for taller buildings more detailed differentiation has to be taken into account. The key to the reference height, $z_e$ for the pressure coefficients, $c_{p_e}$, is given in Fig. 5.2.

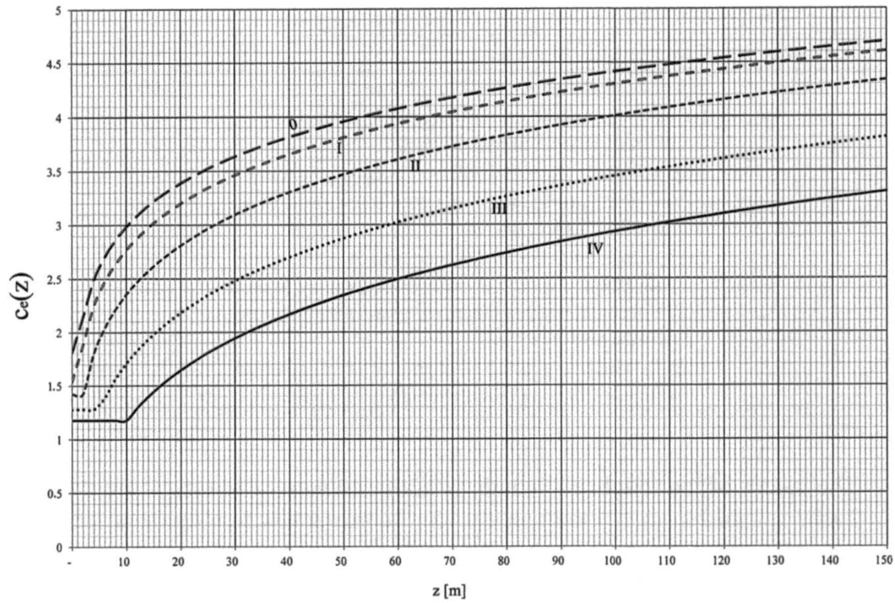


**Fig. 5.1** Exposure factor $c_e(z)$ for terrain roughness and turbulence factors equal to one according Eurocode 1 [16]

**Fig. 5.2** Reference height $z_e$, depending on $h$ and $b$

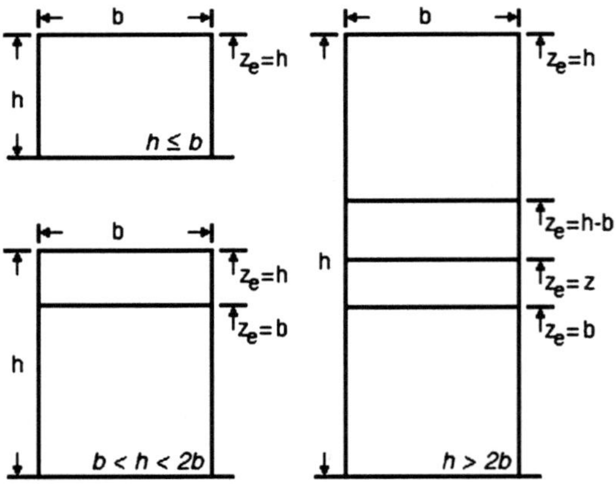


The external pressure coefficients $c_{p_e}$ for buildings and parts of buildings depend on the size of the loaded area, A, which is the area of the element that takes the wind action to be calculated. The part 1–4 of Eurocode 1 provides external pressure coefficients given for loaded areas, A, of 1 and 10 m$^2$ for different building configurations as $c_{p_e,1}$ for local coefficients, and $c_{p_e,10}$, for overall coefficients, respectively.

For loaded areas, A, between 1 and 10 m$^2$ the variation of the values may be obtained from Expressions (5.14) as follows:

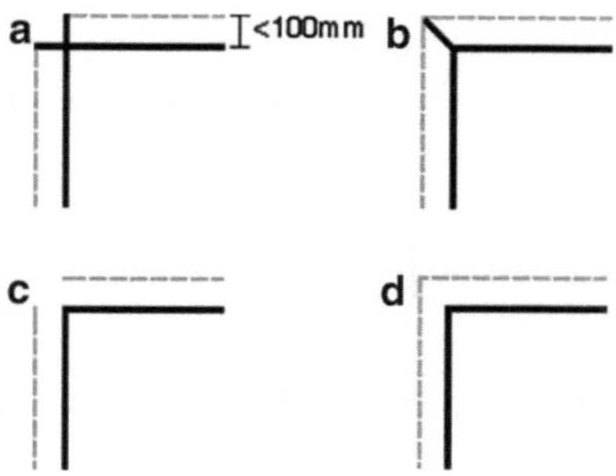


**Fig. 5.3** Corner details for curtain walls with extremities of the cladding skin closed (**a**) and (**b**); Extremities of the cladding skin open (**c**) and (**d**)

$$
\begin{aligned}
c_{pe} &= c_{pe,1}; && A \leq 1\ \mathrm{m}^2 \\
c_{pe} &= c_{pe,1} - (c_{pe,1} - c_{pe,10}) \cdot \log_{10}A; && 1\ \mathrm{m}^2 < A < 10\ \mathrm{m}^2 \\
c_{pe} &= c_{pe,10} && A \geq 10\ \mathrm{m}^2
\end{aligned}
\tag{5.14}
$$

In general the effects of wind friction are only considered on the cladding's surface when the integral area of all surfaces parallel with the wind is greater than four times the entire area of all external surfaces perpendicular to the wind [16] and in these cases merely the strength of the anchorages devices will need to be checked.

#### 5.3.2.1 Internal Pressure

The value of internal pressure of a building is set to by the balance of flow trough openings, into the building, driven by the distribution of external pressure.

Internal pressures for cladding elements shall be considered in ventilated or pressure equalized rainscreen façade systems, *i.e.*, when the joints between cladding's slabs are not sealed.

In fact, the pressure inside the air gap between the cladding and the substructure may vary depending essentially on the permeability, the volume of the air that is enclosed between the skins, the flexibility of the cladding slabs and the air tightness of the interior skin.

The overall pressure difference across two layers may be determined directly from the external and internal pressures. A skin is said to be impermeable when the permeability, defined as the ratio of the total area of openings to the total area of skin, is less than 0.1 %. Caution has to be taken not to confuse with the ratio $\mu$ from Eq. 7.4 of the Eurocode 1.1.4.

If the two skins are impermeable, like in a cavity wall for example, the pressure of the volume of air trapped between them will depend on the atmospheric pressure and temperature. However, when the volume of the air and the stiffness of the skins are small, the net difference temperature can be assumed to be shared equally between the two skins.

If the extremities of the layer between the skins are air tight, which means that the cavity must be effectively sealed along all corners and eaves (Fig. 5.3), and if the

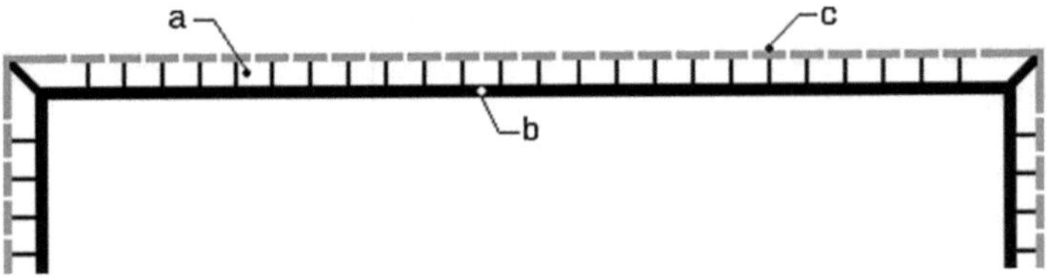


**Fig. 5.4** Pressure equalized rainscreen wall main components: (**a**) PEC – compartment air chamber; (**b**) air barrier system; (**c**) cladding skin with openings

free distance between the skins is less than 100 mm including the thermal insulation material thickness, some internal air pressurization is assumed to occur so that the net internal air pressure on the back of the claddings contributes to a lower value of wind pressure on the cladding.

The following rules may be used, assuming in general, that outside skin includes approximately uniformly distributed openings [16]:

$$c_{p_e,\text{net}} = \frac{2}{3} c_{p_e} \text{ (for external overpressure situations);}$$
$$c_{p_e,\text{net}} = \frac{1}{3} c_{p_e} \text{ (for external underpressure situations).} \tag{5.15}$$

When full pressure equalization is achieved, the outside air pressure is transferred to the air space behind the exterior cladding. In this case the cladding is exposed to a near-zero pressure differential. The rainscreen assembly must comprise three components (Fig. 5.4): the rainscreen itself i.e., the cladding, a compartmented air chamber (PEC) and an air barrier system.

The air chamber compartments must be small enough, the air barrier system must be sufficiently airtight, and the area of the venting through the rainscreen must be large enough to allow sufficient air to move in and out of the compartments under the applied air pressure.

To put it briefly, the goal is to control the airflow within and through the wall assembly to achieve pressure equalization between the outside and the cavity, i.e., zero air pressure differential at all times across the rainscreen.

A complete elimination of the driving force for pressure-induced water penetration [18] is practically impossible, however, perfect pressure equalization across the rainscreen at all times is not achievable either, yet the net pressure acting on the cladding, $c_{p_e,\text{net}}$, may be calculated assuming that the level of pressurization inside the chamber is known.

With the defining of a coefficient of self pressurization, $c_{\text{spc}}$, the ratio between the net pressure in the cladding, $w_{\text{clad}}$, and the external wind pressure, $w_{\text{e}}$, one has:

$$c_{\text{spc}} = \frac{w_{\text{clad}}}{w_{\text{e}}} \tag{5.16}$$

$$c_{p_e,\text{net}} = c_{\text{spc}} \cdot c_{p_e} \tag{5.17}$$

Parametric studies have revealed that values of $c_{spc}$ as lower as 10 % may be achieved in PEC rainscreen systems while in vented systems, where the joints are sealed and only weeps for venting are provided, this coefficient is practically equal to one [19].

In vented systems it's crucial to provide an effective vapour barrier to prevent condensations in the interior side of the stone. The water can become entrapped in kerfs and holes or other cuts, and with freeze-thaw cycling failures will occur due to the loss of the integrity of the stone at the anchorage points. The condensate water will lead to the deterioration of the joint's sealants and to the corrosion of steel. Even supposing that no structural damage occurs, the designer has to be aware that condensate can travel throughout the thickness of the stone and cause loss of strength and staining with aesthetical negative consequences.

#### 5.3.2.2 Application Example

Two buildings, under the same exposure conditions, are considered. A small building 15 m tall and the other 60 m tall, equivalent to the height of 5 and 20 storeys height respectively. Both building have curtain wall façade systems. The tall building's curtain walls have a closed cladding skin system (see Fig. 5.3a) composed of $3 \times 1$ m$^2$ panels and the lower building cladding skin is open (see Fig. 5.3c) with $1 \times 1$ m$^2$ dimension stone cladding slabs.

Both buildings are placed near the sea coast thus the basic velocity (30 m/s) and the orography factor $c_0(z)$ taken as 1.0 are the same. For simplicity external pressure coefficients are $c_{pe,1} = +1.0$ and $c_{pe,10} = +0.8$ are considered in both buildings.

For the lower building the terrain roughness $c_r(z)$ is:

$$c_r(z) = k_r \cdot \ln\left(\frac{z}{z_0}\right) = 0.19 \times \left(\frac{0.003}{0.05}\right)^{0.07} \cdot \ln\left(\frac{15}{0.003}\right) = 1.329;\ \text{Eq.(5.9)}.$$

The mean wind velocity is given by Eq. (5.8):

$$v_m(z) = c_r(z) \cdot c_0(z) \cdot v_b = 1.329 \times 30 = 39.87\ \text{m/s}$$

Considering Eq. (5.7) for the basic pressure:
$q_b = \frac{1}{2}\rho \cdot v_b^2 = 562.5$ Pa and applying Eqs. (5.11) and (5.12) we obtain:

$$q_p = c_e(z) \times q_b = (c_r(z))^2 \left[1 + \frac{7}{\ln(z/z_0)}\right] \cdot q_b = 1810\ \text{Pa} = 1.81\ \text{kPa}.$$

According to Eq. (5.13) one has:

$$w_e = q_p(z_e) \cdot c_{p_e} = 1.81 \times 1 = 1.81\ \text{kPa}.$$

For the taller building and using the same procedure one may find:

$$c_r(z) = k_r \cdot \ln\left(\frac{z}{z_0}\right) = 0.19 \times \left(\frac{0.003}{0.05}\right)^{0.07} \times \ln\left(\frac{60}{0.003}\right) = 1.5453$$

and:

$$q_p = c_e(z) \times q_b = (c_r(z))^2 \left[1 + \frac{7}{\ln(60/0.003)}\right] \cdot 562.5 = 2293 \text{ Pa} = 2.293 \text{ kPa}.$$

Finally and given the rainscreen closed system with 3 $m^2$ panels:

$$c_{p_e} = 1 - (1 - 0.8)\log_{10}3 = 0.905$$

$$w_e = q_p(z_e) \cdot \frac{2}{3} c_{p_e} = 2.293 \times \frac{2}{3} \times 0.905 = 1.383 \text{ kPa}.$$

The reduction for overpressure conditions (and even more for underpressure situations) is clearly relevant when comparing the air pressure obtained for the short building using an open cladding skin system.

### 5.3.3 *Effect of Seismic Action on Cladding*

Severe earthquake causes damage to both structural and non-structural elements such as window glass and curtain walls. In addition the fact that damage endured by curtain wall façades is very costly, falling façade fragments from dimension stone or other any other cladding can pose serious safety hazards even death to both pedestrians and people attempting to leave the building.

In fact, seismic action brings out specific problems not only to designers but to practitioners as well, mainly, due to the lack of rules or regulations about this issue. Design codes tend to limit out-of-plane damage by specifying a seismic static load while damage due to in-plane vibration is controlled by imposing inter-storey drift limits to buildings [20].

During an earthquake two types of lateral loads are considered acting in the façade panels or claddings: the "in plane" loads and the "out of plane" loads. In-plane actions causes mainly shear stresses, while the others excite the panel in bending. The frequency content of the dynamic loads transmitted to the panels is modulated by the building natural frequency, so if by chance it has a value very close to the panel's natural frequency, resonant effects occur with an agonisingly increase of the dynamic response, a well-known phenomena that must be avoided, otherwise structural safety may be compromised.

A simplified method to assess the seismic forces, adapted from the studies of Singh et al. [21, 22], is presented in the following sections.

The characteristic seismic force, $F_{Ek}$, in a given panel or slab depends on the building's natural vibration period, the panel's mass and natural vibration period and the spectral acceleration is evaluated according to Eurocode 8 [23].

$$F_{Ek} = \frac{0.40 \times C_Z \cdot S_{DS} \cdot \gamma_E \cdot M_E}{R_E} \tag{5.18}$$

with, $C_Z$, the acceleration coefficient of the panel, $S_{DS}$, the ground acceleration value, $\gamma_E$, the importance coefficient of the panel (ranging between 1.0 and 1.5), $M_E$, the panel's mass and, $R_E$, a behaviour factor, ranging between 1.5 and 3.5 depending on the panel's performance.

A distinction is made depending on period of natural vibration, $T_{el}$, of the element or panel. A panel is considered to be rigid if its period is less than 0.06 s, which is the same of a frequency in its first mode to be greater than 16.7 Hz, otherwise the element is said to be flexible.

Based on simplified methods, the first mode of the natural frequency of uniformly rectangular slabs with different boundary conditions may be calculated.

For point-supported slabs, a two-way action is mobilized due to the applied transverse loads, and the first mode of vibration can be compared to that of a one-way strip. For a long plate strip resting freely on two opposite sides the period of natural vibration may be expressed as follows [24]:

$$T_{el} = \frac{2L^2}{\pi}\sqrt{\frac{M_E}{EI}} \tag{5.19}$$

with $L$ the span length, $M_E$, the mass per unit length, and, $I$, the modulus of elasticity.

Two different approaches are required depending on the rigidity or flexibility of the element differing on the way that the seismic coefficient from Eq. (5.18) is computed.

Calculating these coefficients it is necessary to know even by estimation the building period, $T_{bd}$. Simple formulas are thus provided depending on the number of floors and the type of the building's structure may be used.

For moment-resisting frame buildings:

$$\begin{aligned} T_{bd} &= 0.054h; \quad \text{(a)} \\ T_{bd} &= 0.034h; \quad \text{(b)} \\ T_{bd} &= 0.025h; \quad \text{(c)} \end{aligned} \tag{5.20}$$

Expression (a), (b) and (c) of Eq. (5.20) corresponds to bared frames, infilled frames with openings and fully infilled frames respectively [25].

For concrete shear wall buildings the following formula may be used [26]:

$$T_{bd} = \frac{0.09h}{\sqrt{B}} \tag{5.21}$$

In Eqs. (5.20) and (5.21), $h$, and, $B$, are the height and the dimension of the building at its base in the direction under consideration expressed in meters.

Two methods are thus presented to the evaluation of the corresponding two acceleration coefficients: $C_{Rz}$, for rigid elements and, $C_{Fz}$, for the flexible elements.

#### 5.3.3.1 Rigid Elements

An element is said to be rigid if its estimated period is less than 0.06 s. A distinction is made between short and tall buildings. A short building is classified when it has eight floors or less and a tall building when the number of floors, $N$, is greater than eight.

For short buildings the rigid element acceleration coefficient, $C_{Rz}$, is defined:

$$C_{Rz} = 1 + \frac{z}{h}(C_N - 1) \tag{5.22}$$

For tall buildings ($N > 8$) the rigid element acceleration coefficient may be found using the following expressions:

$$\begin{aligned}
&C_{Rz} = 1 + \tfrac{z}{h_1}(C_1 - 1); && z \le h_1 = 0.2 \times h && \text{(a)}\\
&C_{Rz} = C_1; && h_1 < z \le h_2 = 0.8 \times h && \text{(b)}\\
&C_{Rz} = C_1 + \frac{z - h_2}{h - h_1}\cdot(C_N - C_1); && z > h_2 && \text{(c)}
\end{aligned} \tag{5.23}$$

where:

$z$ is the height of the element in the building ground level;
$h$ is the height of the building;

$C_1$ an acceleration coefficient value for intermediate floors which is defined in terms of the building period $T_{bd}$:

$$C_1 = \frac{C_N}{\sqrt[3]{T_{bd}}} \tag{5.24}$$

$C_N$ an acceleration coefficient at the roof level according to the following Equation.

$$C_N = b\sqrt{1 + 1.03 \times R_1^2} \ge 1 \tag{5.25}$$

with $b$ the product of the first mode participation factor with the first mode shape value at the roof level and $R_1^2$ is the normalized spectral response acceleration at the fundamental period.

In order to evaluate Expression (5.25) one may assume a linear variation of the first mode and the following approximate expression, depending on the number of building stories, $N$, can be used to define, $b$:

$$b = \frac{3 \times N}{2 \times N + 1} \tag{5.26}$$

$R_1$ is defined in Eq. (5.27), as a function of, $S_a$, the spectral response acceleration at the fundamental period, and $S_{DS}$ the design spectral response at a short period, i.e., of the terrain ground:

$$R_1 = \frac{S_a}{0.4 \times S_{DS}} \tag{5.27}$$

#### 5.3.3.2 Flexible Elements

As in the case of the rigid elements first the floor acceleration is calculated by the methods described in the previous section and then it is amplified by a factor that depends on the periods of the building.

After determination of the building period (see Eqs. (5.20) and (5.21)) the formulae for the flexible acceleration coefficient, $C_{Fz}$ of an element depends on the value of the element's period of vibration, $T_{el}$.

$$C_{Fz} = C_{Rz} \cdot A \geq 1.0 \tag{5.28}$$

with $C_{Rz}$ the acceleration coefficient for a rigid component as defined in Eqs. (5.22) and (5.23), and $A$ an amplification factor which values depend on the value of the element period:

$$\begin{aligned} &A = 1; && T_{el} < 0.06 && \text{(a)} \\ &A = 1 + \frac{T_{el} - 0.06}{T_m - 0.06}(A_m - 1); && 0.06 \leq T_{el} < T_m && \text{(b)} \\ &A = A_m; && T_m \leq T_{el} < T_{bd} && \text{(c)} \\ &A = A_m e^{-2,5\times(T_{el} - T_{bd})} \geq 1; && T_{el} \geq T_{bd} && \text{(d)} \end{aligned} \tag{5.29}$$

$A_m$ is an amplification factor that depends upon the periods of the building and on the critical damp factor of the element, $\xi$, and is given by the following Equations:

$$\begin{aligned} &4 \leq A_m = \frac{7}{T_{bd}^{0.3}} \leq 9; && \xi = 2\% \\ &3 \leq A_m = \frac{4.6}{T_{bd}^{0.3}} \leq 6; && \xi = 5\% \end{aligned} \tag{5.30}$$

$T_m$ in Eq. (5.29) is a period that depends on the building period $T_{bd}$, the number of storeys $N$, and the floor number $m$ where the element is placed, defined as follows:

$$T_m = \frac{T_{bd}}{2 \times (N - m) + \sqrt{N}} \tag{5.31}$$

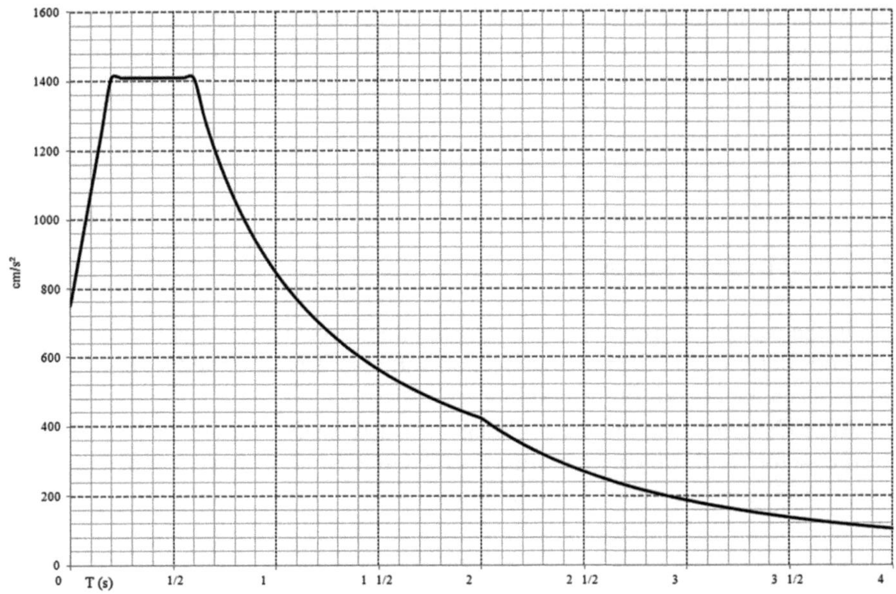


**Fig. 5.5** Horizontal acceleration response spectrum Type 1 for 5% damping and a subsoil class C according Eurocode 8

#### 5.3.3.3 Application Examples

Two application examples are presented regarding the determination of the induced forces in two cladding slabs subject to a seismic action which horizontal acceleration response spectrum is defined in Fig. 5.5.

Rigid Elements

Let us consider at first a building with fully infilled resisting moment frames with 20 floors, 3 m height each.

The cladding consist of $1 \times 1\ \text{m}^2$ panels, 0.05 m thick granite slabs with a volumic mass of 2,500 Kg/m$^3$ and a modulus of elasticity of 30 GPa.

The period of a cladding slab is

$T_{el} = \frac{2L^2}{\pi}\sqrt{\frac{M_E}{EI}} = \frac{2\times 1^2}{\pi}\sqrt{\frac{125\times 12}{30\times 10^9\times 0.05^3}} = 0.0127\ \text{s} < 0.06\ \text{s}$, confirming that the element is rigid according to the implicit classification criteria.

For the 20 storey building (60 m height) the fundamental period is given (see Eq. (5.20) thus $T_{bd} = 0.025h = 0.025 \times 60 = 1.5$ s.

The acceleration coefficient is calculated according the respective formulae and for the most critical situation one has Eq. (5.23):

$R_1 = \frac{S_a}{0.4 \times S_{DS}} = \frac{563.5}{0.4 \times 751.33} = 1.875$, from Eq. (5.27);
$b = \frac{3 \times N}{2 \times N+1} = \frac{3 \times 20}{2 \times 20+1} \simeq 1.463$, from Eq. (5.26);
$C_N = b\sqrt{1 + 1.03 \times R_1^2} \simeq 3.146 \geq 1$, from Eq. (5.25) and
$C_1 = \frac{C_N}{T_{bd}^{1/3}} \simeq 2.748$ from Eq. (5.24).

Thus to a slab attached to the building's façade at the middle of the last floor $z = 58.5\,\text{m} > 0.8 \times 45 = 36\,\text{m}$ (see expression (5.23c)) the acceleration coefficient is:

$$C_{Rz} = C_1 + \frac{z - h_2}{h - h_1} \cdot (C_N - C_1) = C_1 + \frac{58.5 - 0.8 \times 60}{60 - 0.2 \times 60} \cdot (C_N - C_1) \approx 2.835$$

Assuming the importance coefficient with unit value and the behaviour factor $R_E = 1.5$, the characteristic seismic force, $F_{Ek}$, induced in the slab in the above mentioned conditions is then calculated according Expression (5.18).

$$F_{Ek} = \frac{0.40 \cdot C_{Rz} \cdot S_{DS} \cdot \gamma_E \cdot M_E}{R_E} \simeq 710\ \text{N}.$$

To compare the force that will be installed in the same cladding slab under the same seismic conditions yet in a lower building, the calculations are now adjusted to a same structure type building, yet with 5 storeys only.

In these circumstances the necessary adjustment are to be made:

For the lower building (15 m height) the fundamental period is given (see Eq. (5.20) thus $T_{bd} = 0.025h = 0.025 \times 15 = 0.375$ s.

This is a short building ($N \leq 8$) the rigid element acceleration coefficient is defined:

$R_1 = \frac{S_a}{0.4 \times S_{DS}} = \frac{1408.75}{0.4 \times 751.33} \simeq 4.688 \rightarrow$ Eq. (5.27);
$b = \frac{3 \times N}{2 \times N+1} = \frac{3 \times 5}{2 \times 5+1} \simeq 1.364 \rightarrow$ Eq. (5.26);
$C_N = b\sqrt{1 + 1.03 \times R_1^2} \simeq 6.629 \geq 1 \rightarrow$ Eq. (5.25).

Considering the same situation of a slab attached to the building's façade at the middle of the last floor $z = 13.5$ m

$C_{Rz} = 1 + \frac{z}{h}(C_N - 1) = 1 + \frac{13.5}{15} \times (C_N - 1) \simeq 6.066 \rightarrow$ Eq. (5.22).

Assuming the same parameter values for the importance and behaviour factors the characteristic seismic force, $F_{Ek}$, induced in the slab in a short building is then calculated according Expression (5.18).

$$F_{Ek} = \frac{0.40 \cdot C_{Rz} \cdot S_{DS} \cdot \gamma_E \cdot M_E}{R_E} \simeq 1\,519\ \text{N}.$$

Comparing the values obtained in the tall building with the one in the short building the ratio between the induced the forces in the $1 \times 1 \times 0.05$ m$^3$ cladding slabs is practically equal to two, corresponding to a uniform distributed horizontal load of 0.71 and 1.52 kPa respectively.

Flexible Elements

Let us consider the same two buildings with a different type of cladding slabs.

The cladding consist of $1 \times 3$ m$^2$ panels, 0.06 m with a volumic mass of 2,200 Kg/m$^3$ and a modulus of elasticity of 30 GPa.

The period of a cladding slab is:

$T_{el} = \frac{2L^2}{\pi}\sqrt{\frac{M_E}{EI}} = \frac{2\times 3^2}{\pi}\sqrt{\frac{132\times 12}{30\times 10^9 \times 0.06^3}} = 0.0896$ s $> 0.06$ s, confirming that the panel is flexible.

According to Eq. (5.31) and considering the element in the last floor $m = 20$, we obtain:

$$T_m = \frac{T_{bd}}{2 \times (N - m) + \sqrt{N}} = \frac{1.5}{\sqrt{20}} \simeq 0.335 \text{ s}$$

Considering a critical damp factor equal to 5% the value of the amplification factor $A_m$ (Eq. (5.30) is for the tall building:

$$A_m = \frac{4.6}{T_{bd}^{0.3}} = \frac{4.6}{1.5^{0.3}} \simeq 4.073$$

For the 20 storey building the amplification factor given by Eq. (5.29)b, $A = 1 + \frac{T_{el} - 0.06}{T_m - 0.06}(A_m - 1) \approx 1.33$, and finally the flexible acceleration coefficient, $C_{Fz}$ is calculated according to Eq. (5.28):

$$C_{Fz} = C_{Rz} \cdot A = 2.835 \times 1.33 \simeq 3.771$$

Finally

$$F_{Ek} = \frac{0.40 \cdot C_{Fz} \cdot S_{DS} \cdot \gamma_E \cdot M_E}{R_E} \approx \frac{0.4 \times 3.771 \times 751.3/100 \times 1 \times 132}{1.5}$$

$$\approx 997.3 \text{ N}$$

For the lower building the fundamental period is $T_{bd} = 0.375$ s and the period of the element is $T_{el} = 0.0896$ s.

For the last floor $T_m = 0.168$ s.

Considering the same value for the critical damp factor equal to 5 % (Eq. (5.30)) the value of the amplification factor $A_m$ is:

$$A_m = \frac{4.6}{T_{bd}^{0.3}} \leq 6 \rightarrow A_m = 6$$

and

$$A = 1 + \frac{T_{el} - 0.06}{T_m - 0.06}(A_m - 1); \quad 0.06 \leq T_{el} < T_m$$

$$A = 1 + \frac{T_{el} - 0.06}{T_m - 0.06}(6 - 1) \simeq 2.373$$

$$C_{Fz} = C_{Rz}A \approx 6.066 \times 2.373 \approx 14.396$$

and finally:

$$F_{Ek} = \frac{0.40 \cdot C_{Fz} \cdot S_{DS} \cdot \gamma_E \cdot M_E}{R_E} \simeq 3\,807 \text{ N}$$

Comparing the values obtained in the tall building with the short building it may be observed that the induced forces in the $1 \times 3 \times 0.06\text{m}^3$ panels are practically the same corresponding to a uniform distributed horizontal load of 0.332 and 1.269 kPa respectively.

### 5.3.4 Thermo-Hygrometric Action

The expansion and contraction of materials are mainly dependent on environmental thermo-hygrometric variations and on the stone's properties.

It is the responsibility of the designer to provide the location and type of vertical and horizontal movement joints in any façade wall system not only to accommodate movement due to thermal expansion and contraction of the claddings, but also those due to creep and shrinkage of concrete structural elements, deflection of supporting structures, drying shrinkage of wood frame and earthquake movements.

Movement joint spacing for veneer depends to some degree on the rigidity of the support system. Structural steel typically will have larger deflections than concrete frame buildings. Veneers on taller buildings also require horizontal movement joints formed by gaps under shelf angles to accommodate vertical movement.

In this section some guidance is given for the calculation of the joint's spacing width regarding movements originate by thermo-hygrometric actions.

**Table 5.4** Indicative values – hygrometric coefficient of expansion [10]

| Natural stone | Average values $\alpha_m$ |
|---|---|
| Granite | 0.010 – 0.0200 |
| Limestone | 0.008 – 0.025 |
| Marble | 0.005 – 0.010 |

#### 5.3.4.1 Thermal Induced Movements

Temperature variation in a construction element when under sun rays has much bigger proportions than those due to the air temperature oscillation.

In sun oriented surfaces during in the summer the faced cladding temperature can reach very high values especially in case of dark tone varieties.

In some regions and circumstances over a 24 h period the temperature amplitude may range more than 80 °C.

The thermal dimensional change, $\Delta_{lt}$, may be estimated with the simple formula:

$$\Delta_{tl} = \alpha_t \cdot \Delta_t \cdot L \tag{5.32}$$

with $L$ the joints spacing length, $\alpha_t$, the coefficient of thermal expansion and $\Delta_t$, the temperature variation of the element. Some indicative values of the coefficient of thermal expansion are depicted on Table 2.1 in [mm/mK].

#### 5.3.4.2 Hygrometric Movements

Stone, like any material expands and contracts in relation to its moisture content which depends on the environment and exposure conditions.

In order to estimate the dimensional variations due to hygrometric conditions the hygrometric coefficient of expansion, $\alpha_m$, is used. It expresses the variation of a length in percentage per percent of the relative humidity variation.

In Table 5.4 typical values are given for the most used natural stones.

Denoting again the spacing between joints, $L$, and by $\Delta_m$ the variation in percentage of the moisture content and $\alpha_m$ the hygrometric coefficient of expansion, the thermal dimensional change due to moisture may be determined using the following expression:

$$\Delta_{ml} = \alpha_m \cdot \Delta_m \cdot L \tag{5.33}$$

For example a cladding in marble façade with a coefficient of 0.010 % per percent of change in relative humidity restrained between two columns spaced by 50 m with an increase in relative humidity of from 10 to 50 % will have an unrestrained linear expansion of:

$$\Delta_{ml} = \alpha_m \cdot \Delta_m \cdot L = 0.01\% \times (50 - 10)\% \times 50{,}000 = 2 \text{ mm}.$$

### 5.3.5 Snow Loading

Extensive activity research measured data of many meteorological stations worldwide and the characteristic values of the ground snow load were determined by means of extreme value statistics.

Moreover, functions were developed which allow to describe the interdependency between the characteristic value of the ground snow load and the height of the relevant site above sea level. For example, the new European snow load map as well as the relationship between altitude and snow load is given in EN 1991-1-3 [27].

Snow load is only considered in horizontal or in small slope surfaces such as some roofs, balconies, terraces or verandas. If mechanically fixed dimension stone is used as roof or pavement cladding, the snow load and eventually wind overpressures, have to be combined with the slabs self-weight.

A roof shape coefficient depending on the roof angle together with an exposure and thermal coefficient are used to define the snow load for a particular region.

The general formula given is [27]:

$$s = \mu_i \cdot C_e \cdot C_t \cdot s_k \tag{5.34}$$

where $\mu_i$ is the roof shape coefficient, $C_e$ the exposure coefficient, $C_t$ the thermal coefficient and $s_k$ the characteristic value of the ground snow load for the relevant altitude and region.

### 5.3.6 Induced Vibrations

In some locations, vibrations induced by heavy traffic to the buildings structure foundations are transmitted to façades and its cladding, compromising a satisfactory performance especially when cracking, even invisible, is present.

Thus, it is indispensable to gather special procedures to take into consideration these situations. For example, slots and holes should be filled with resilient material to ensure that contact between the anchorage devices is uniform, damping the induced vibrations.

## 5.4 Dimension Stone Strength

Manufacturing and fabrication of natural stone products, such as dimension stone cladding has developed in such a rapid way, that technical and scientific knowledge did not encompass it with the appropriate terms of use, particularly in respect to the criteria for verification of safety and design.

In the last decades, enormous development of the cutting technologies suddenly brought to stone cladding façade's designers and suppliers the build capabilities of extraordinary reduced thickness.

Meanwhile, the scientific development has granted a much better understanding of the material's behaviour and then the possibility of the use of modern design tools. For instances, limit state design to cladding stone can and may be applied in the same manner that is presently employed for other "man-made" materials.

Because natural stone is used "as it is", disregarding post cutting treatments, its physical and mechanical properties have to be gathered expressly for a suitable volume of work either in square meters of façade cladding or in a certain volume of extracted stone in quarry. Due to the randomness of the strength's properties of some types of natural stone it is necessary to conduct tests in a greater number, than in the case of "man-made" materials.

As far as the dimension stone cladding is concerned it's the flexural and tensile strengths that are crucial to its performance and ability as a cladding material.

In some specific situations like the anchorage areas of prestressing devices the compression strength of the stone may become critical.

In the following sections, information and guidance is given to obtain the characteristic values of dimension stone strength based on or derived from the information that is collected in some of the relevant standard tests.

### *5.4.1 Tests and Data Analysis*

Thickness of stone veneer on buildings has been significantly reduced. One of the obvious consequences is the huge number of substantial failures that arise from using thin dimension stone cladding without base on evidence and sound research.

The samples to be tested should be drawn from normal production blocks, extracted from the same quarry defined for a given project. This is to say, that historical data only gives an indication on strength in an informational relative basis about the stone to be used.

Designers must be aware that stone is anisotropic. Its properties are directionally dependent.

The method of deposition in sedimentary rocks, in which the layers or beds are predominantly orientated or bedded, or in the case of metaphoric rocks, sedimentary transformed by heat and pressure or, even in the case of igneous rocks, due to the cooling of the structure, which may cause micro predominant cracking in a particular direction or directions. This formation phenomenon gives rise to anisotropic behaviour.

The stone appearance in various directions will ask for a decision on the preferred orientation. Thus, the aesthetical effect may have a strong influence on the mechanical relevant properties and then on the cost. This is a key issue to consider during initial testing [28].

**Table 5.5** Quantile factor, $k_{p,5\%}$, in dependence on the number of measured values, $n$, and the distribution skewness, for a confidence level $\gamma = 0.75$

| n | 3 | 4 | 5 | 6 | 7 | 8 | 9 | 10 | 15 | 20 | 30 | 40 | 50 | $\infty$ |
|---|---|---|---|---|---|---|---|---|---|---|---|---|---|---|
| $\alpha = 0$ | 3.15 | 2.68 | 2.46 | 2.34 | 2.25 | 2.19 | 2.14 | 2.10 | 1.99 | 1.93 | 1.87 | 1.83 | 1.81 | 1.64 |
| $\alpha = -1$ | 4.31 | 3.58 | 3.22 | 3.00 | 2.86 | 2.76 | 2.69 | 2.63 | 2.45 | 2.33 | 2.23 | 2.19 | 2.09 | 1.85 |

On the other hand, the designer must be aware that different surface textural finishes may affect the properties of the stone, the reason why every strength test must be performed with the samples exhibiting the same surface treatment that will be used on the building's façade.

### 5.4.2 *Characteristic Values*

The pull-out strength, flexure strength, the compression and tensile strength of the stone are fundamental to determine the structural reliability and thus a statistical evaluation of the test results that have to be performed for design calculations.

In this case the strength is a significant variable in the limit state verification represented by characteristic values, which correspond to a prescribed probability of not being infringed. Usually the lower value of a material property or product is unfavourable and the 5 % lower fractile is then considered as the characteristic value.

Following the assumptions that were taken for the ultimate limit state and the definition of the partial factors, a log-normal distribution is considered to approximate the data from the tests (see Sect. 4.4.3). As so a lower characteristic value, $R_k$, is obtained as follows:

$$R_k = e^{\bar{x}_{\ln} - \left(k_{p,5\%} \times s_{\ln}\right)} \tag{5.35}$$

where, $\bar{x}_{\ln}$ is the logarithmic mean and $s_{\ln}$, the standard deviation of the $n$ tested specimen is given by:

$$\bar{x}_{\ln} = \frac{1}{n} \sum_{j} \ln\left(x_j\right); \quad s_{\ln} = \pm \sqrt{\frac{\sum \left(\ln\, x_j - \bar{x}_{\ln}\right)^2}{n-1}} \tag{5.36}$$

The definition of the coefficient $k_{p,5\%}$, depends on the number of measured values, $n$, as well on the confidence level, $\gamma$, in correspondence to the 5 % fractile.

To take account of statistical uncertainty the value, $\gamma$, of 0.75 is recommended in ISO 2394 [29]. Therefore the corresponding coefficients for $k_{p,5\%}$ are given in Table 5.5. The values are given depending on the skewness of the distribution of the sample.

When the number of the samples is small the asymmetry of the population distribution may have a significant effect essentially when there is high variability and small probabilities, which often is the case with natural stone. In these situations it may be convenient to calculate the skewness $\alpha$ of the distribution sample and interpolate the values of $k_{p,5\%}$ accordingly.

$$\alpha = \frac{n}{(n-1)(n-2)} \times \sum_j \left( \frac{(\ln x_j) - \bar{x}_{\ln}}{s_{\ln}} \right)^3 \tag{5.37}$$

### 5.4.3 *Design Assisted by Testing*

Two very distinct standards are available for the quantification of the breaking load at the dowell hole. In both standards load is applied, separately, perpendicular to the surface of the panel and it's not implied to address all if any safety concerns, being the responsibility of the designer to establish appropriate safety margins and to determine the applicability of the obtained results.

The European standard EN 13364 [30] is not being followed by the European industry because, besides being expensive, its procedures take too much time, the test conditions are very stringent and they are far from representing the real in-site anchorage situations and or conditions, e.g., tests are realized only in dry specimens and a full rigid connection between the dowell and the stone is assumed apparently disregarding any prying action.

The American standard ASTM C1354 [31] is a general purpose test simple to implement for any anchorage system and prudently recommending that the specimens should be soaked.

Test specimens consisting of a stone panel sample and a mechanical anchor are fabricated in the same manner and of the same materials as their intended construction uses. The mechanical anchor is connected to a test support. A test load is applied perpendicular or parallel to the face of the stone panel.

Using the test setup illustrated in Fig. 5.6 load is applied gradually using a calibrated test machine and increased until the stone or the mechanical anchor fails.

The anchorage system shall be representative of the type to be used in field construction and shall include the anchor to be used and all accessories normally required to attach the anchor to the backup structure. The value of the breaking load of the anchorage is evaluated according the simple formula equalling moments (see Fig. 5.6):

$$P = T\frac{g}{L} \tag{5.38}$$

The stone capacity of an anchor is difficult to accurately predict mathematically since the relative stiffness of stone to the anchor, the type of infill material and

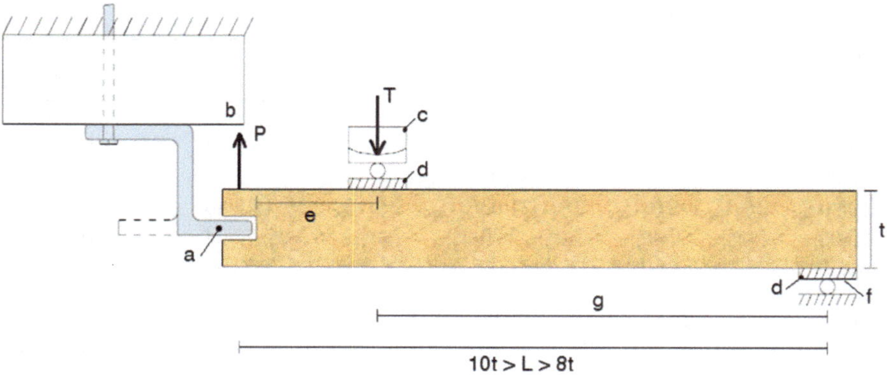


**Fig. 5.6** Setup for testing connection according ASTM C 1354. (**a**) anchor; (**b**) supporting fixture; (**c**) loading rod; (**d**) rubber pads; (**e**) minimum distance of load to the cut in edge equals to thickness of specimen; (**f**) support along full specimen width; (**g**) distance of load to the uncut edge; *L* distance between supports minimum 8 × t and maximum 10 × t; (*T*) test cell load; (*P*) anchorage breaking load

back-up, control how much of the anchorage device is actually resisting. In fact for the correct estimation of the design breaking load it is necessary to assume at least that: the resistance is a function of the set of independent quantities; test number is satisfied; so that the resistance the normal or lognormal distribution is valid and all relevant geometric and material properties are measured.

#### 5.4.3.1 Example of Application

20 samples were used in tests designed to determine the breaking load at the dowell hole in terms of ASTM C1354. The samples, measuring 300 mm × 400 mm, were prepared from slabs of 30 mm nominal thickness.

The holes were located between 98 and 102 mm from either side, measured to the nearest 0.5 mm. The thickness of stone between the edge of the hole and the two faces was $(10 \pm 2)$ mm, measured to the nearest 0.5 mm. The diameter of the holes was $(10 \pm 0.5)$ mm. The depth of the holes was $(30 \pm 2)$ mm.

In Table 5.6 the relevant information is presented. The values of the breaking load at the dowell hole are determined using Eq. (5.38) and the corresponding logarithms are listed in the right-most column.

Using the above values one can find:

$$\bar{x}_{\ln} = \frac{1}{n} \times \sum_j \ln\left(x_j\right) = 0.796257$$

$$s_{\ln} = \pm\sqrt{\frac{\sum\left(\ln x_j - \bar{x}_{\ln}\right)^2}{n-1}} = 0.181893$$

**Table 5.6** Pull-out test data according ASTM C1354

| L [mm] supports distance | g [mm] distance to cut edge | t [mm] specimen thickness | T [kN] Test load | P [kN] Breaking load | ln (P) [kN] |
|---|---|---|---|---|---|
| 358.00 | 294.50 | 31.00 | 3.68 | 3.03 | 1.10856 |
| 357.75 | 295.00 | 31.00 | 2.71 | 2.23 | 0.80200 |
| 360.00 | 295.00 | 30.00 | 3.15 | 2.58 | 0.94779 |
| 360.00 | 295.00 | 31.00 | 3.42 | 2.80 | 1.02962 |
| 350.00 | 280.50 | 29.00 | 2.86 | 2.29 | 0.82855 |
| 350.75 | 284.00 | 31.00 | 2.84 | 2.30 | 0.83291 |
| 348.75 | 282.50 | 31.00 | 2.03 | 1.64 | 0.49470 |
| 350.00 | 282.00 | 31.00 | 2.56 | 2.06 | 0.72271 |
| 350.50 | 283.50 | 30.50 | 2.39 | 1.93 | 0.65752 |
| 346.75 | 281.50 | 30.50 | 1.99 | 1.62 | 0.48243 |
| 350.00 | 282.00 | 31.00 | 2.80 | 2.26 | 0.81536 |
| 349.25 | 283.00 | 31.00 | 2.93 | 2.37 | 0.86289 |
| 350.25 | 279.50 | 30.50 | 2.57 | 2.05 | 0.71784 |
| 345.00 | 279.50 | 30.50 | 3.36 | 2.72 | 1.00063 |
| 344.00 | 244.50 | 31.00 | 4.05 | 2.88 | 1.05779 |
| 344.50 | 251.50 | 31.00 | 3.16 | 2.31 | 0.83725 |
| 347.50 | 279.00 | 31.50 | 2.61 | 2.10 | 0.74194 |
| 353.50 | 287.00 | 31.00 | 2.08 | 1.69 | 0.52473 |
| 350.00 | 277.00 | 30.50 | 2.96 | 2.34 | 0.85015 |
| 346.25 | 276.50 | 30.00 | 2.31 | 1.84 | 0.60977 |

The skewness $\alpha$ of the distribution sample is given by:

$$\alpha = \frac{n}{(n-1)(n-2)} \times \sum_j \left( \frac{(\ln x_j) - \bar{x}_{\ln}}{s_{\ln}} \right)^3 =$$

$$\alpha = \frac{20}{19 \times 18} \times (-2.46952) = -0.1444$$

To obtain the characteristic value of the stone the data of Table 5.5 is used. Interpolation gives the value of $k_{p,5\%} = 1.9877(7)$ and applied in Eq. (5.35):

$$R_k = e^{\bar{x}_{\ln} - \left(k_{p,5\%} \times s_{\ln}\right)} = e^{0.796257-(1.9877 \times 0.181893)} = 1.5444 \text{ kN}$$

Considering that the coefficient of variation of the sample is 16.4 % the design value of breaking load at the dowell is obtained as follows:

$$\sigma_{Rd} = \eta \cdot \frac{\sigma_{Rk}}{\gamma_M} = \frac{1.5444}{2.4} \approx 0.64354 \text{ kN}$$

with, $\gamma_M$, the partial safety factor according Table 4.3 and assuming an aging factor equal to 1.

# References

1. CEN – European Committee for Standardization (2004) Eurocode 0: basis of structural design, vol EN 1990. CEN, Brussels
2. British Standards Institute (2010) Code of practice for the design and installation of natural stone cladding and lining. Traditional handset external cladding, vol BS 8298–2:2010. BSI, London
3. Materials A-ASfTa (2010) Standard guide for selection, design, and installation of dimension stone attachment systems, vol ASTM C1242 – 10. American Standard of Testing Materials, West Conshohocken
4. Materials A-ASfTa (2010) Standard guide for selection of dimension stone for exterior use, vol ASTM C1528 – 10. American Standard of Testing Materials, West Conshohocken
5. Gere AS (1988) Design considerations for using stone veneer on high-rise buildings. B. Donaldson edn. ASTM – American Society for Testing Materials, Philadelphia
6. Lewis MD (1995) Modern stone cladding: design and installation of exterior dimension stone systems, vol MNL 21, ASTM manual series. ASTM – American Society for Testing Materials, Philadelphia
7. Marble Institute of America (2007) Dimension stone design manual – VII. Marble Institute of America Cleveland, Cleveland
8. Camposinhos RS (2009) Revestimentos em Pedra natural com Fixação Mecânica, vol 1. primeira edição edn. Edições Sílabo, Lisboa
9. CEN – European Committee for Standardization (2007) EN 1991-1-1 Eurocode 1: actions on structures — Part 1–1: general actions – densities, self-weight, imposed loads for buildings. CEN TC 250
10. RdS C (2009) Revestimentos em Pedra natural com Fixação Mecânica – Dimensionamento e Projecto, vol 1, 1st edn. Edições Sílabo, Lisboa
11. Lungu D, Trandafir R, Gelder Pv (1996) Comparative study of Eurocode 1, ISO and ASCE procedures for calculating wind loads. International Association for Bridge and Structural Engineering (IABSE), Delft, Netherlands
12. Cook NJ, Establishment BR (1985) The designer's guide to wind loading of building structures. Butterworths [for] Building Research Establishment, Department of the Environment
13. Mehta KC (1998) Wind load standards. In: Jubileum conference on wind effects on buildings and structures, Porto Alegre, Brazil
14. Holmes JD (2007) Wind loading of structures, 2nd edn. Taylor & Francis, London/New York
15. Overend M, Zammit K (2006) Wind loading on cladding and glazed façades. Paper presented at the international symposium on the application of architectural glass, Munich
16. CEN – European Committee for Standardization (2007) EN 1991-1-4 Eurocode 1: actions on structures — Part 1–4: general actions – wind. CEN TC 250
17. CEN – European Committee for Standardization (2002) Eurocode 1: actions on structures — Part 1–4: general actions – wind actions. CEN TC 250
18. Rousseau MZ, Poirier GF, Brown WC (1998) Pressure equalization in rainscreen wall systems. Construction technology updates, vol 17. Institute for Research in Construction, Ottawa
19. Camposinhos RS (2005) Fachadas-cortina pressurizáveis. [S. n.], Porto
20. SucuoÇğlu H, Vallabhan CVG (1997) Behaviour of window glass panels during earthquakes. Eng Struct 19(8):685–694
21. Singh MP, Moreschi LM, Suárez LE, Matheu EE (2006) Seismic design forces. I: rigid non-structural components. J Struct Eng 132(10):1524–1532. doi:10.1061/(asce)0733-9445(2006)132:10(1524)
22. Singh MP, Moreschi LM, Suárez LE, Matheu EE (2006) Seismic design forces. II: flexible non-structural components. J Struct Eng 132(10):1533–1542. doi:10.1061/(asce)0733-9445(2006)132:10(1533)
23. CEN (2010) Eurocode 8 – design of structures for earthquake resistance. Part 1: general rules, seismic actions and rules for buildings, vol NP EN1998-1. IPQ, Caparica

24. Szilard R (1974) Theory and analysis of plates: classical and numerical methods. Prentice-Hall civil engineering and engineering mechanics series. Prentice-Hall, Englewood Cliffs
25. Crowley H, Pinho R (2006) Simplified equations for estimating the period of vibration of existing buildings. Paper presented at the first European conference on earthquake engineering and seismology, Geneva, Switzerland
26. Goel RK, Chopra AK (1998) Period formulas for concrete shear wall buildings. J Struct Eng 124(4):426–433. doi:10.1061/(ASCE)0733-9445(1998)124:4(426)
27. CEN – European Committee for Standardization (2003) Eurocode 1 – actions on structures – Part 1–3: general actions – snow loads. CEN TC 250
28. Camposinhos RS, Pires V, Amaral P, Rosa LG (2011) Slate flexural and anchorage strength considerations in cladding design. Constr Build Mater 25(10):3966–3971. doi:10.1016/j.conbuildmat.2011.04.029
29. Engineering MSL (1999) Technical review of ISO 2394 general principles on reliabilities for structures. Health & Safety Executive, Bootle
30. CEN – European Committee for Standardization (2002) EN 13364. Determination of the breaking load at the dowel hole. Natural stone. Test methods. CEN – TC 246, Brussels
31. Testing ASf, Materials (2004) C1354-06: standard test method for strength of individual stone anchorages in dimension stone. ASTM

# Chapter 6
# Dowel Anchorage

**Abstract** Installation procedures and design assumptions are presented to the dowel anchorage system. Holes location on slabs edges are discussed taking into consideration the induced stresses by flexure under lateral actions.

Minimum slabs thickness formulae are also presented depending on the flexural strength capacity or pull-out strength capacity are provided, taking into consideration stress concentrations either in the mid-span or in the support region.

A formula to find maximum sag due to deformation under gravity load is presented.

An application example to determine minimum thickness is presented for common and several slabs dimensions.

## 6.1 Introduction

Suitable to slabs with length sizes less than about 1 m, anchorage systems making use of insertions of dowel pins into drilled holes are one of the most used anchorage systems for dimension stone cladding.

The fixing itself is provided by four dowel pins inserted into cylindrical holes at half thickness on two parallel slab's edges. After installation the holes can be positioned in horizontal or vertical position.

Each one of the parallel edges contains two holes properly spaced with an insertion depth of about 2/3 of the slab's thickness depending on the dowel's diameter.

The dowels work together with a perpendicular rod or shaft, normally a spade bolt with a lock nut to fix in the body anchor or in a tubular section grout-in anchor as illustrated in Fig. 6.1.

With body anchors the distance to the natural stone panel is adjustable by the spade bolt with the possibility to swivel the anchor laterally a considerable amount, thus responding on-site, to the required adjustability.

R. de Sousa Camposinhos, *Stone Cladding Engineering*,
DOI 10.1007/978-94-007-6848-2_6, © Springer Science+Business Media Dordrecht 2014

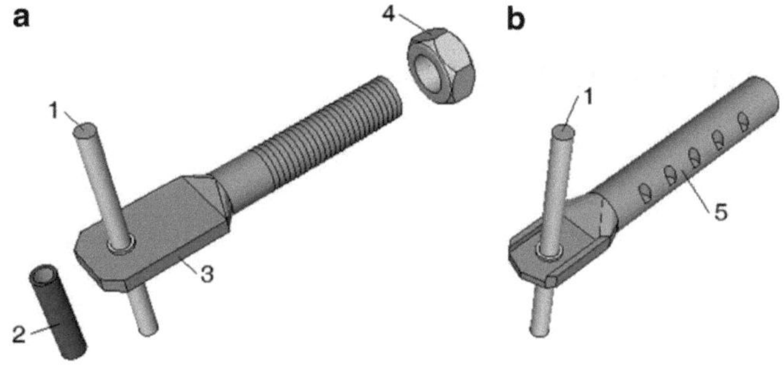


**Fig. 6.1** Typical dowel anchor systems: (**a**) to be attached to supporting devices; (**b**) traditional grout-in method system; (*1*) loose pin; (*2*) sleeve; (*3*) rod in a spade bold format; (*4*) lock nut; (*5*) tubular grout-in anchor

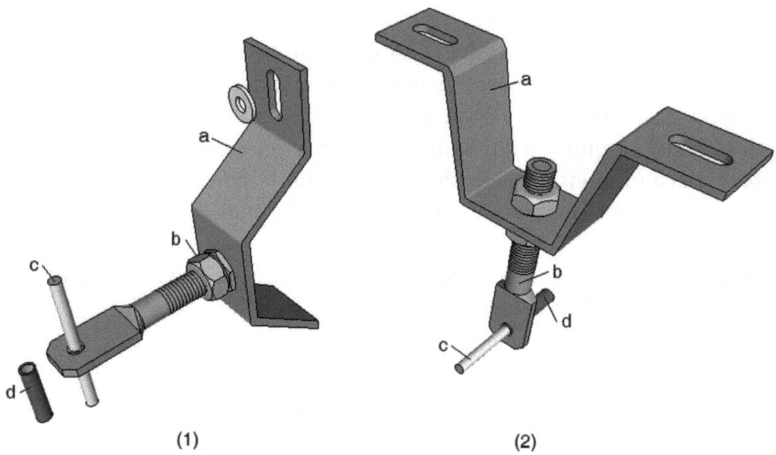


**Fig. 6.2** Anchor body types: (*1*) for use in vertical or horizontal joints, (*2*) for fixing natural slabs in horizontal areas; legend: (*a*) body anchor; (*b*) lock nut; (*c*) loose pin; (*d*) sleeve

Body anchors connections can be made to several different load-bearing sub-structures, not only in façade vertical surfaces but also in horizontal surfaces (Fig. 6.2).

## 6.2 Design Procedures and Assumptions

As previously referred, attention should be given to the number and arrangement of rods. There are always four for each panel, two per edge for edges to be mounted horizontally or vertically. Each rod engages with one or two dowels or pins.

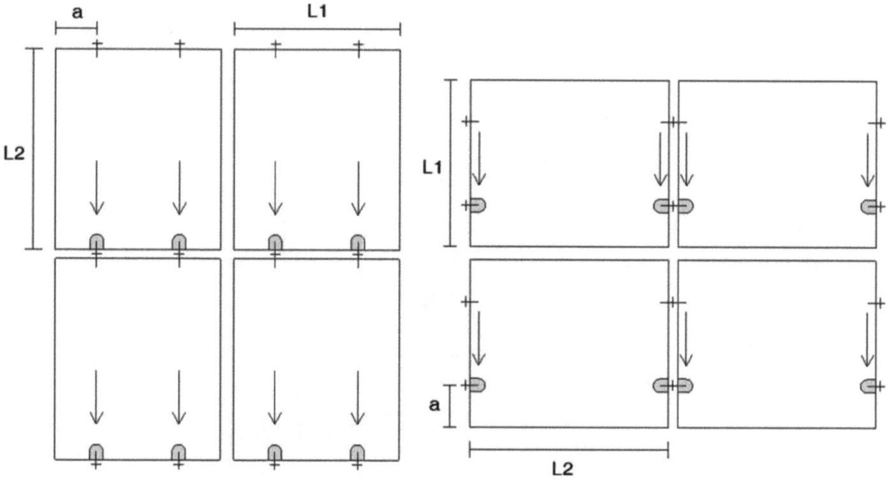


**Fig. 6.3** Panel configuration and dimensions, dowel number per edge, location and self-weight

Self-weight is to be supported only by lower anchors whether on the vertical or horizontal edges as well as the lower pins when anchoring is performed on vertical edges. The lateral loads are supported evenly by all dowels and rods.

That is why the lower rods need to be more resistant because they must bear the whole of a plate's self-weight. It is recommended that there are no more than two holes per edge so as to avoid overstress states resulting from the misalignment of holes [1].

A schematic representation of both configurations is illustrated in Fig. 6.3. In the figure each arrow represents half self-weight of one panel. The distance from the centre of a hole to the closest edge is denoted by, $a$, $L_1$, and $L_2$, are respectively the dimensions of the supported and free edges.

Dowel pin common diameter and length are about 6–7 mm and 30–60 mm respectively.

Each dowel pin is engaged in the drilled hole which is approximately 2–3 mm larger than the diameter to allow the space for a sleeve.

A typical cross section is illustrated in Fig. 6.4 noting that for horizontal joints anchors can support gravity and lateral actions as well (figure as section view), for anchors in vertical joints only lateral actions are transmitted.

The main issues relating to the mounting of stone panels using dowel anchors revolve around considerations that are implicit in the design which include, among others and relating to the same Fig. 6.4, minimizing the eccentric weight (e) on the anchorage and preventing any slip of the connection after vertical adjustment with a diagonally welded slotted washer plate or serrated anchor and washer (f), using a horizontal slot to align dowels with holes and vertical slot for height adjustment (d), inserting plastic or metal adjustment bearing shims that are slightly larger than

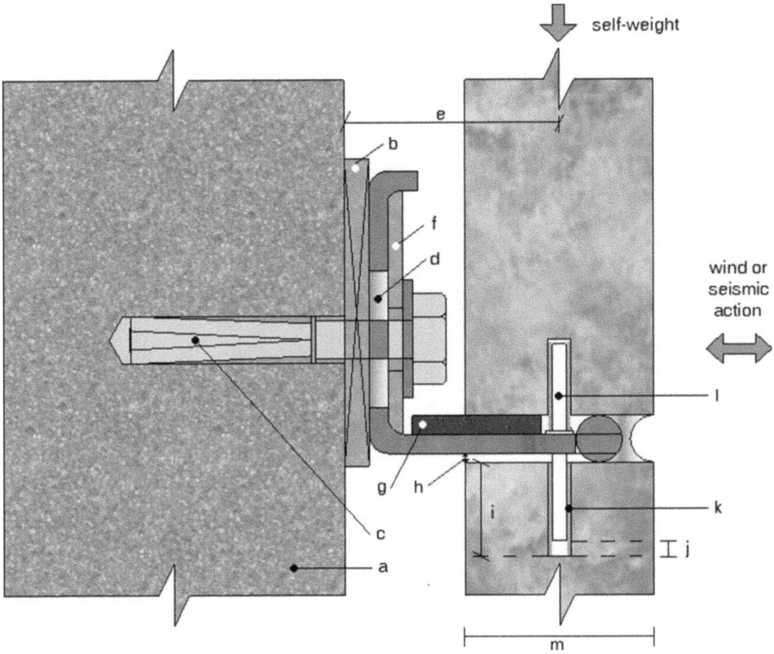


**Fig. 6.4** Typical cross-section of dowel anchorage systems. (*a*) backup structure; (*b*) adjusting shim; (*c*) fastener to backup; (*d*) slots in body anchor for vertical and horizontal adjustment; (*e*) eccentricity of the stone weight; (*f*) serrated or welded washer; (*g*) resilient bearing material; (*h*) body anchor clearance; (*i*) hole depth; (*j*) clearance to avoid point loading on dowel ends; (*k*) sleeve; (*l*) dowel pin; (*m*) hole location in centre third of slab thickness

the anchor's fastened face (b), inserting plastic or stiff rubber shims for levelling and separation (g), locating holes in the central third of panel thickness (m), maintaining clearances to avoid point loading on the dowel end (j), to avoid dowel misalignment with the hole and resulting prying on the stone panel [2–4].

For structural stone plate design, the cladding system needs to be considered as a separate structure. Configuration and reaction of the supports have to be suitably and fully defined. In this case, the structural model that best represents the situation is the one way slab system. The point of fixture of the metallic insert and the slab is considered, from a structural point of view, to be a simple support, which, by definition, prevents any translational movement perpendicular to the slab's surface.

The main operational load on the cladding comes from wind or earthquakes, which exert bending. Other applied loads such as self-weight and imposed deformations are not usually taken into consideration, as their effect is not relevant to the stress state of vertically placed panels. Any imposed deformations are generally considered to be engaged by the space between the insert and the slab. To estimate slab dimensions, the considerations below need to be taken into account upon definition of lateral actions.

### 6.2.1 *Flexure Design*

In order to assure that actions are transmitted from the slabs to the supports the drilled holes have to be filled with resilient material to assure that an even repartition of bearing between all contact points is granted, otherwise a three-point supporting system may be formed for which the assumptions in the design are not compatible.

In that manner resilient sleeves will also accommodate some misalignment in the drilled holes and will absorb the induced vibrations either by the wind or by heavy traffic in the close proximity of the buildings. For the purposes of optimizing panel capacity, the distance of the hole, $a$, (see Fig. 6.3) from the unsupported edge can be defined by assuming that the maximum positive and negative bending moments have the same absolute value and Eq. (6.1) may be established.

$$a = \frac{L_1}{4} \cdot \frac{\sqrt{2}}{1 + \sqrt{2}/2} \approx 0.21 \times L_1 \tag{6.1}$$

If the supported length, $L_1$, is greater than the unsupported length, $L_2$, it is possible to establish an expression for the limit value for the ratio of the panel dimensions to define which direction is governing the maximum bending stresses on the panel.

$$\frac{L_2}{L_1} < \frac{1}{1 + \sqrt{2}} \approx 0.41 \tag{6.2}$$

If the conditions in Eqs. (6.1) and (6.2) are both satisfied, the designer may assume that the governing bending moment is given by Eq. (6.3):

$$M_{Sd} = \frac{q_{Sd} \cdot L_2^2}{8} \tag{6.3}$$

where, $M_{Sd}$, is the design value of applied internal bending moment, and, $q_{Sd}$, is the design value of the lateral load.

To take in account the stress concentration along the supports region the value of the ratio given in Eq. (6.2) needs to be adjusted accordingly [5].

Thus, considering the notation used in Fig. 6.5, if, $L_2/L_1 > 0.5$, the governing bending moment (M3) is parallel to $L_2$, otherwise if, $L_2/L_1 < 0.5$, the governing bending moment (M2 or M1) is parallel to $L_1$, provided that Eq. (6.1) is satisfied.

It has to be pointed out that in some conditions, the stereotomy of the façade may dictate a different condition of that defined by Eq. (6.1).

For example, when the designer pretends to define the joint's alignment in one direction in such a manner to intercept the middle length of the joints spacing in the other direction, different ratios have to be defined. However, it must be emphasised that in the majority of situations, the smaller side is seldom inferior to half the dimension of the longer size and, in general, flexure design is governed by the bending moment along the smaller dimension provided that the value of the dowel's insertion is defined accordingly the above-mentioned conditions.

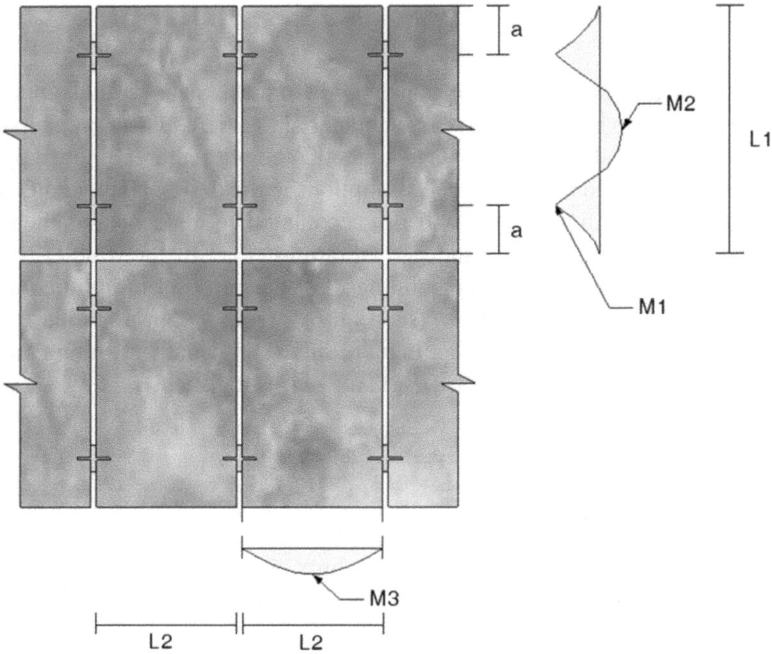


**Fig. 6.5** Panels dimension proportion according to the distance from the drill to the edge and the panel side's dimension

#### 6.2.1.1 Ultimate Limit State

As previously mentioned, in most cases the governing bending moment is parallel to L2 direction (see Fig. 6.5) and the maximum bending stress is located at mid span along a band strip connecting the supports. This happens due to the non-uniform distribution of the bending moment values across a section parallel to $L1$ (see Fig. 6.6). In this way stress concentrations have to be regarded when computing the maximum flexure tensile stress.

Stress concentration factors are obtained, in general, from the elasticity theory using analytical solutions [6], experimental analysis [7, 8] or the finite element method. These factors represent the ratio of the maximum stress, $\sigma_{\max}$, in some region or cross section to the reference stress taken in this case as a medium stress, $\sigma_{\text{med}}$, as represented in Fig. 6.6.

Consequently, the design value of the maximum bending stress is expressed in Eq. (6.4):

$$\sigma_{Sd} = \frac{3 \cdot q_{Sd} \cdot L_2^2 \cdot k_{d2}}{4 \cdot t^2} \tag{6.4}$$

with the same notation of Eqs. (6.2) and (6.3) with $k_{d2}$ the stress concentration factor for the, $L_2$, direction and, $t$, the nominal thickness of the slab.

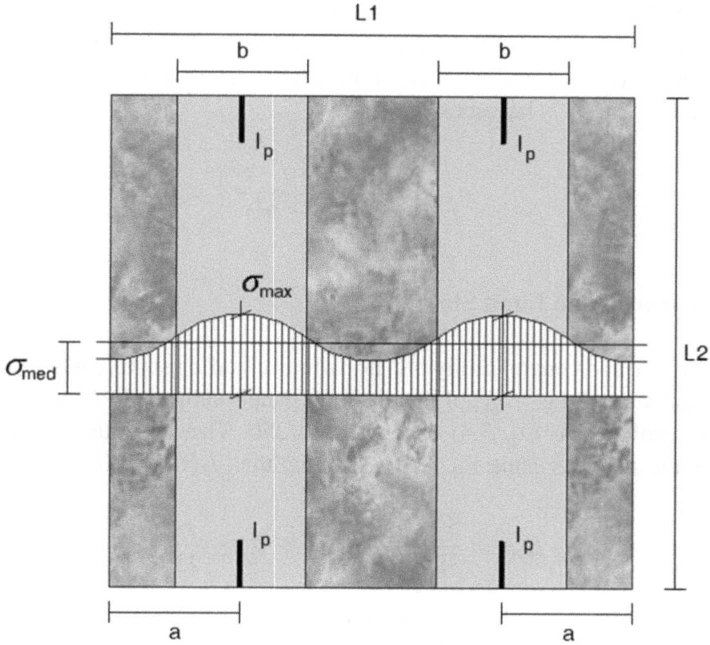


**Fig. 6.6** Typical slabs support configuration, dowel's embedment depth, $l_p$, and concentration of stresses due to bending along the mid-cross section; a, distance of dowels to nearest edge; b, idealized strip width; $L_2$, free edge length; $L_1$, supporting edge length

The stress concentration factor, $k_{d2}$, depends on several factors, such as the dowel pin length and the characteristic of the material's sleeve. For typical supporting conditions, $k_{d2}$, may be taken with the value of 1.30 [5] and thus Eq. (6.4) is simplified:

$$\sigma_{Sd} = \frac{0.975 \times q_{Sd} \cdot L_2^2}{t^2} \tag{6.5}$$

The design value of the stone flexure strength, $\sigma_{Rd}$, being known, the thickness of the dimension stone can be estimated. Assuming that $\sigma_{Sd} \leq \sigma_{Rd}$, it results that:

$$t \geq L_2\sqrt{\frac{0.975 \times q_{Sd}}{\sigma_{Rd}}} \approx L_2\sqrt{\frac{q_{Sd}}{\sigma_{Rd}}} \tag{6.6}$$

For the less common cases where $L_2/L_1 < 0.5$ the direction of the governing bending moment ($M_2$ or $M_1$ in Fig. 6.5) is parallel to, $L_1$, and the acting maximum bending stress will occur in the drilled which is calculated as follows:

$$\sigma_{Sd} = \frac{3 \times q_{Sd} \cdot a^2 \cdot k_{d1} \cdot t}{(t^3 - \phi^3)} \tag{6.7}$$

with $a$, the distance of the hole axis to the nearest slab's edge, $k_{d1}$, the stress concentration factor taking into account the geometry variation of the supports, and, $\phi$, the diameter of the hole. In general it may be assumed that, $k_{d1}$, takes the value of 1.5 [5], thus Eq. (6.7) becomes:

$$\sigma_{Sd} = \frac{9 \times q_{Sd} \cdot a^2 \cdot t}{2 \times (t^3 - \phi^3)} \tag{6.8}$$

#### 6.2.1.2 Deformation Limit State

The appearance and general utility of a horizontal cladding surface may be impaired when the calculated sag, $f_{\max}$, of the cladding panels subjected to quasi-permanent load combination (see Eq. 5.4) exceeds span/250. The sag must be assessed in relation to the supports using Eq. (6.9) assuming that, $L_2/L_1 \geq 0.5$.

$$f_{\max} = \frac{5 \times q_{Sd} \cdot L_2^4}{384 \times E \cdot I}(1 + \psi) \leq \frac{L_2}{250} \tag{6.9}$$

where, $L_2$, is the length of the unsupported span, $E$, the elastic modulus of the stone, $I$, is the moment of inertia of the slabs cross section per unit width, $q_{Sd}$, is the design value of the quasi-permanent load combination acting on the slab (see Table 5.2) and, $\psi$, the creep coefficient of the type of stone which value may be taken as two [9].

### *6.2.2 Pull-Out Strength Design*

It has been pointed out that for the general panels dimensions in the vicinity of the supports that a greater stress concentration occurs, thus the stone thickness is more dependent on the pull-out strength design than in the flexural design. This is the reason why the focus on the bending strength of stone panels to the detriment of anchorage design has been pointed out as being the cause of a significant number of failures [1, 10].

Prediction of the anchorage capacity of the dowel pin in drilled hole mathematically requires the validation tests to be made.

There are several difficulties involving the modelling of the problem of this task. The dowel anchor capacity is limited by several parameters such as the dowel hole's diameter, the distance from the edge of the stone to the edge of the hole, the distance at which the dowel pin is inserted into the drilled hole, the depth that the dowel actually contacts the stone which is directly related with the prying action that always occurs. In fact, the relative stiffness of the slab, anchor device and backup control the way how the anchorage device is resisting the action transmitted to the whole assembly.

**Fig. 6.7** Cone type failure after a pull out test of a dowel in a sleeved hole in a marble specimen exhibiting a partial contact length.

The area around the hole depth in contact with the dowel pin either through a rigid connection or through a sleeve's resilient material will determine the type of the contact between the dowel and the hole.

There will be either a point contact, at the rim of the hole, or along an unpredictable length along its root (Fig. 6.7).

On the other hand the stone strength at the stone's affected area is crucial to determine the actual breaking load.

Studies that have been made involving several pull-out tests [1, 5, 11] and the relationships between material strength, anchorage strength, and induced stress states for dowel anchorage and material property tests were examined. The findings on several stone types conclude that the projected area of the stone failure surface for an edge dowel can be reasonably approximated by a triangle. An idealized post-break aspect of the stone is illustrated in Fig. 6.8.

The maximum principal stress theory can be applied satisfactorily to a limiting normal stress. Failure occurs when the normal stress reaches a specified upper limit. Based on this assumption, and the depicted failure area and angle, the value of the breaking load at the dowel hole can be estimated with sufficient accuracy.

The use of the finite element method with linear-elastic material properties, together with the maximum principal stresses failure criteria when module of rupture is attained, has been shown to be an appropriate design procedure for estimating the breaking load at the dowel hole [1, 12].

Stress concentrations are present in the vicinity of the hole and quickly dissipate, suggesting that edge dowel anchorage failures are influenced by a non-uniform stress distribution over the failure surface. The direction of the maximum principal stresses located adjacent to the hole are approximately perpendicular to the failure surface observed in tests, meaning that failure initiates at this location (Fig. 6.9).

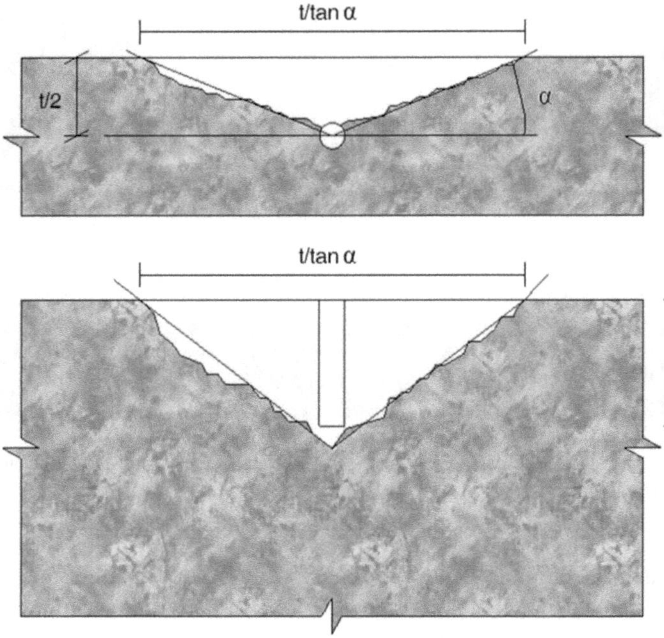


**Fig. 6.8** Idealized failure surface section and plane view upon a pull out test of a dowel in drilled hole

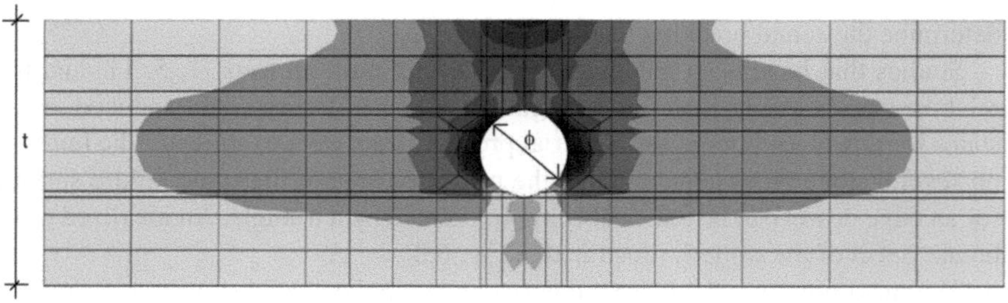


**Fig. 6.9** Stress maps 2D view on drilled face obtained from a finite element analysis showing stress concentration with the darker zones meaning higher tensile stress

### 6.2.2.1 Simplified Formula

Let us consider a simple case where a perfect rigid connection between the dowel and the hole is assumed as depicted in Fig. 6.10.

The detaching spall is initiated at the drilled hole location, where the horizontal diameter crosses the hole wall. Lines drawn at an angle α from the horizontal assumed initiation location to the stone edge define the spall width which length is depicted in Fig. 6.8.

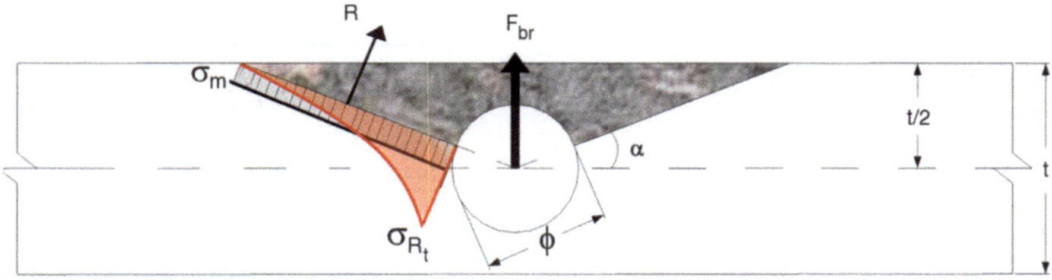


**Fig. 6.10** Section view for representing stress distribution and resultants in the drilled edge of a dimension stone

Considering that $R$ represents the resultant force of the stresses installed in each half of the spall its value can be expressed as follows:

$$R = A \times \frac{\sigma_{Rt}}{K_1} \tag{6.10}$$

where, $A$, is half of the projected horizontal surface area of the spall, $\sigma_R$, is the stone tensile strength and, $K_1$, is a stress concentration factor defined by the ratio between the actual maximum stress, $\sigma_{Rt}$, and the medium stress, $\sigma_m$, according to Fig. 6.10.

The breaking load value, $F_{br}$, can be estimated using the formula:

$$F_{br} = 2 \times R \cos \alpha \tag{6.11}$$

The maximum principal stress theory underpins this formulation as, in fact, it is satisfactorily applicable to brittle or quasi-brittle materials, such as stone. The theory is based on a limiting normal stress. Failure occurs when the normal stress reaches a specified upper limit and may be predicted when either of the principal tensile or compression stresses equals or exceeds the ultimate strength, $\sigma_{Rt}$, of the material.

It must be emphasized, as mentioned in Chap. 2, that the type of finishing for the stone surface significantly alters the resistance of the panels in terms of either the bending strength or the breaking load at the dowel hole.

Empirical studies allowed determining the value of, $\alpha$, in different conditions and, referring to Figs. 6.8 and 6.10, the value of half of the projected horizontal surface area of the spall, $A$, is given by:

$$A = \frac{1}{4}\left(\frac{t}{\sin \alpha} - \phi\right) \times l_d \tag{6.12}$$

where, $\phi$, is the drilled hole diameter, $t$, the thickness of the dimension stone, $l_d$, the contact depth of the pin dowel and, $\alpha$, is the spall angle from primary surface slab's plane. As assumed, the calculated area from Eq. (6.12) considers a simplified pyramid shape for the spall with the same height of the contact length of the dowel.

**Table 6.1** Stress concentration factor $K_2$ and spall angle $\alpha$ to be used in Eq. (6.14)

| Stone Type | $\alpha$ | $K_2$ |
|---|---|---|
| Fine grained granite | 30° | 1.2 |
| Oolitic limestone | 21° | 1.3 |
| Crystalline marble | 19° | 1.4 |
| Semi-crystalline limestone | 22° | 1.6 |

Substituting Eqs. (6.12) and (6.10), in Eq. (6.11) the breaking load at the dowel hole can be estimated as follows:

$$F_{br} = \frac{\sigma_{Rt} \cdot l_d \,(t - \phi \times \sin\alpha)}{2 \times K_1 \times \tan\alpha} \tag{6.13}$$

When a resilient material is used as a sleeve, such as EPDM rubber, the stress concentration is higher than the one found in a rigid connection [5, 12] so that an additional stress concentration factor is applied to take into account the resulting prying effect [5] and Eq. (6.13) has to be adjusted.

On the other hand the value of $K_1$ has been found to be approximately equal to 3 for the case of cylindrical holes [1, 5, 13, 14]. Thus, considering the characteristic tensile strength value of the stone and the aging factor previously discussed in Chap. 3, to estimate the breaking load design value at the dowel hole, $F_{brd}$, the following equation is established:

$$F_{brd} = \frac{\sigma_{Rtk} \times l_d \times (t - \phi \times \sin\alpha)}{6 \times K_2 \times \tan\alpha} \times \frac{\eta}{\gamma_M} \tag{6.14}$$

with

$\sigma_{Rtk}$ – characteristic tensile strength value of the stone;
$t$ – thickness of the dimension stone;
$\phi$ – drilled hole diameter;
$l_d$ – contact depth of the pin dowel;
$\alpha$ – spall angle from primary surface slab's plane;
$K_2$ – concentration factor to take into account the additional prying action induced by a resilient sleeve;
$\gamma_M$ – partial factor of safety for the stone, according to Table 4.3;
$\eta$ – aging factor taking into account the loss of strength of the stone according to Table 4.4.

The values of the stress concentration factors, $K_2$, presented in Table 6.1 correspond to the effect of the resulting prying action and consequent stress amplification due the presence of a resilient sleeve between the dowel and the stone. In the same table the corresponding average values of the spall angles, $\alpha$, are presented according the author studies [1, 13, 15].

#### 6.2.2.2 Design Procedure

The design value of the action or load transmitted to a dowel anchorage may be obtained through the following equation:

$$F_{Sd} = \frac{w_{Sd} \cdot L_1 \cdot L_2}{4} \tag{6.15}$$

where, $w_{Sd}$, is the design value of the uniform the lateral load acting in a dimension stone slab and $L_1$, $L_2$, are the size lengths of the slab.

The safety verification format implies that $F_{brd} \geq F_{Sd}$ allowing for the determination of the minimum thickness stone:

$$t \geq 1.5 \times w_{Sd} \cdot L_1 \cdot L_2 \cdot K_2 \frac{\tan \alpha}{\sigma_{Rtd} \cdot l_d} + \phi \times \sin \alpha \tag{6.16}$$

with $\sigma_{Rtd} = \frac{\sigma_{Rkt}}{\gamma_M} \cdot \eta$ the design value of the tensile strength of the stone.

## 6.3 Application Example

Fine grained granite dimension stone slabs are under consideration to be used as a façade cladding using pin dowel anchorages. The determining lateral action, $w_{Sd}$, was found to have a value of 2 kPa.

Tests on specimens for the determination of flexural strength under constant moment were performed in two different groups of specimens: standard specimens and dimension stone slabs, in order to determine the characteristic value of the tensile strength with Eq. 2.9.

Thus, the design value of the flexural strength and of the tensile strength of the granite was found to be 2.5 and 1.5 MPa respectively. These values were obtained dividing the characteristic value by a partial safety coefficient, $\gamma_M$, with a value of 2.9, according to Table 4.3 upon the calculation of the coefficient of variation of the experimental data with a value of 23 %.

Dimension stone slabs are supported along the longer edge, $L_1$, with the holes positioned as defined by Eq. (6.1). A range of values between 400 and 1,000 mm is considered in the example for the calculation of the required thickness of the slabs following two different criteria: based on the flexural strength and on the pull-out strength by means of Eqs. (6.6) and (6.16) respectively.

For aesthetical reasons a constant ratio value between the slabs size lengths, $L_2$ and $L_1$, is assumed to be equal to 2/3. The relevant data is summarized in Table 6.2.

**Table 6.2** Data used in the relevant formula for the application example

| $w_{sd}$ | $\sigma_{Rd}$ | $\sigma_{Rtd}$ | $\frac{L_2}{L_1}$ | $\alpha$ | $\phi$ | $l_d$ | $K_2$ |
|---|---|---|---|---|---|---|---|
| 2.0 kPa | 2.5 MPa | 1.5 MPa | 2/3 | 32° | 8 mm | 40 mm | 1.2 |

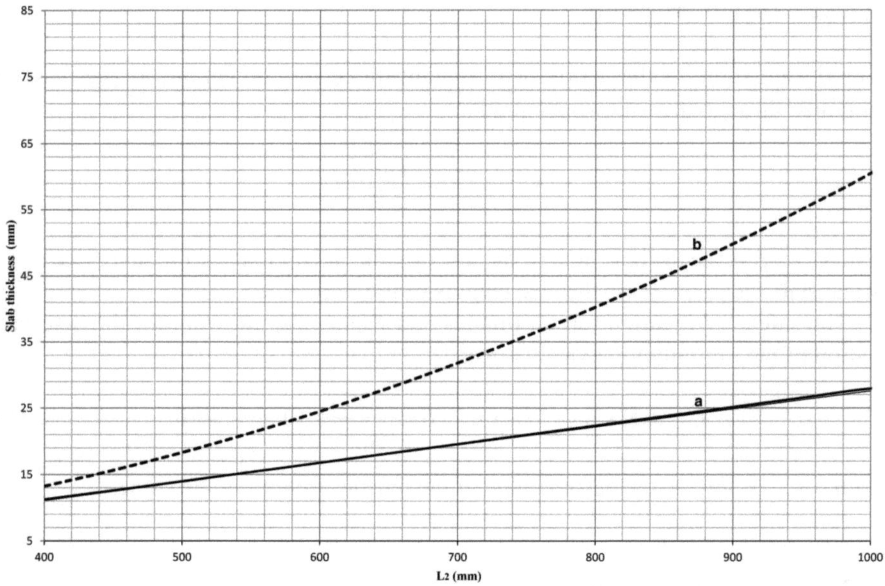


**Fig. 6.11** Required minimum thickness of dimension stone slabs considering the flexural strength (line a) and the pull-out strength (line b) as a function of the span length $L_2$ according to data from Table 6.2

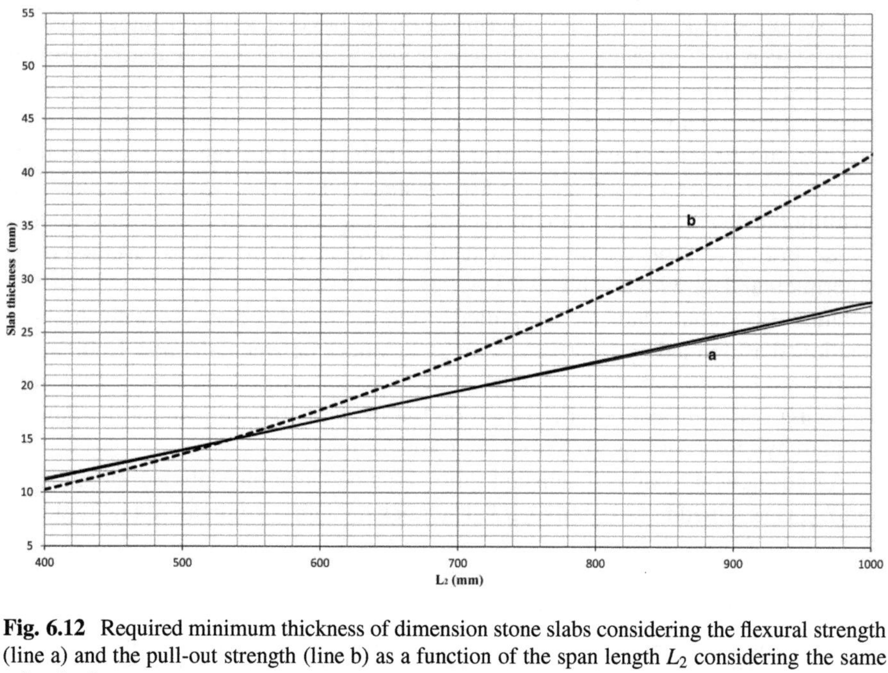


**Fig. 6.12** Required minimum thickness of dimension stone slabs considering the flexural strength (line a) and the pull-out strength (line b) as a function of the span length $L_2$ considering the same value for $L_1$

The required minimum thickness was calculated and the results are presented in graphic form in Fig. 6.11.

The required thickness in the anchorage zone, as a result of the pull-out strength, increases as the span length increases.

This increase is proportional to the increase of the ratio between, $L_2$, and $L_1$.

Interesting to note that in a limit situation, i.e., for squared slabs only for smaller length sizes it could be expected that flexural strength will govern the design. This situation is illustrated in Fig. 6.12 for the same problem but the ratio between $L_2$ and, $L_1$, has a unit value, which corresponds to square slabs.

## References


1. Camposinhos RS, Camposinhos RPA (2009) Dimension-stone cladding design with dowel anchorage. Proc ICE Constr Mater 162(3):95–104. doi:10.1680/coma.2009.162.3.95
2. Materials A-ASfTa (2010) Standard guide for selection, design, and installation of dimension stone attachment systems, vol ASTM C1242 – 10. American Standard of Testing Materials, West Conshohocken
3. Lewis MD (1995) Modern stone cladding: design and installation of exterior dimension stone systems, vol MNL 21, ASTM manual series. ASTM, Philadelphia
4. Marble Institute of America (2007) Dimension stone design manual – VII. Marble Institute of America Cleveland, Cleveland
5. Camposinhos RS (2009) Revestimentos em Pedra natural com Fixação Mecânica – Dimensionamento e Projecto, vol 1, 1st edn. Edições Sílabo, Lisboa
6. Timoshenko S, Goodier JN (1970) Theory of elasticity. Engineering societies monographs, 3rd edn. McGraw-Hill, New York
7. Dally JW, Riley WF (1991) Experimental stress analysis. McGraw-Hill, New York
8. Ákos Rechtorizs IB, Miklós Gálos (1998) Determination of stress intensity factors on rock specimens. Paper presented at the symposium in civil engineering, Budapest
9. Singh DP (1975) A study of creep rocks. Int J Rock Mech Min Sci Geomech Abstr 12:271–276
10. West DG, Heinlein M (2000) Anchorage pullout strength in granite: design and fabrication influences. In: Hoigard KR (ed) Dimension stone cladding. ASTM – American Society for Testing and Materials, West Conshohocken
11. Camposinhos RS (2012) Undercut anchorage in dimension stone cladding. Proc ICE Constr Mater. doi:10.1680/coma.11.00050
12. Vera Pires AP, Infante V, Amaral PM, Rosa LG (2012) Finite element model development for contact analysis of two different dowel fixing conditions of Portuguese granites "Cinzento de Alpalhão" and "Amarelo de Vila Real" in cladding. Paper presented at the Global Stone Congress 2012, Borba, Portugal
13. Oliveira MdFTdS (2008) Revestimentos de fachadas de edifícios em pedra natural fixada mecanicamente. IST, Lisboa
14. Pilkey WD (2008) Stress concentration factors, 3rd edn. Wiley, New York
15. Camposinhos RS (2009) Study on the effect of cylindrical drilling of polyethylene sheaths. Study on the effect of "epdm" strips on continuous support – experimental analysis and computer modelling. Polytechnic of Porto – School of Engineering, Porto

# Chapter 7
# Undercut Anchorage

**Abstract** Installation procedures and design assumptions are presented to the undercut anchorage technology, in particular its behaviour and performance.

Undercut geometry and location on slabs-backing are discussed taking into consideration the induced stresses by lateral actions.

Minimum slabs thickness formulae depending on the flexural strength capacity or pull-out strength capacity are provided.

Special considerations are taken to the stress field induced in the support region through formulae derivation calibrated by experimental studies and finite element method.

A formula to find maximum sag due to deformation under gravity loads is also presented.

An application example to determine minimum thickness is presented for common and several slabs' dimensions and results are discussed.

## 7.1 Introduction

Resistance to lateral loading, mainly wind and seismic action, is usually achieved by means of stainless steel anchors inserted into kerfs or holes that are drilled or cut into the edges of the stone panels. These anchors are mechanically connected to the building's structure, thus providing the essential mechanical connection between the stone and the structure.

One structural weak point in this type of stone construction is to be found at the kerfs or anchor holes on the edges of the stone slabs. These cuts need to leave sufficient stone thickness to provide the necessary strength to resist the various forces or actions to which the stone panels are subject.

Undercut anchoring systems keying type anchors carry the tensile load through main keys at the end of the anchor, resulting in a cone shape failure or in the yielding of the steel rod. This technology, combined with a suitable support framework,

R. de Sousa Camposinhos, *Stone Cladding Engineering*,

DOI 10.1007/978-94-007-6848-2_7, © Springer Science+Business Media Dordrecht 2014

allows the engineer to implement a safe and high performance fixing system for stone cladding façades. Undercut anchoring has proven to be a very efficient system yet it has been commented to be more expensive than other systems.

When heavier panels are used, or where there are no relevant cost constraints, the tragic consequences of an eventual failure may dictate the reason why this anchoring system is preferred when comparing with other usual edge supporting systems.

## 7.2 Undercut Anchoring

Generally fasteners can be subdivided into three different working principles according to the load transfer mechanism, namely friction, bonding and keying.

For friction type anchors, the tensile load is transferred from the anchor to the base material due to the friction created by expanded segments. For bonding anchors, the tensile load is transferred mainly through the adhesive bond between the anchor rod and the stone which may cause a combined shear and cone type failure.

Keying type anchors carry the tensile load through main keys at the end of the anchor, resulting in a cone shape failure or in the yielding of the steel rod. This is the case of undercut anchoring. This technology, combined with a suitable support framework, allows implementing a safe and high performance fixing system for stone cladding façades.

The undercut drilling is performed with a proper drill bit and a special drilling device in order to obtain the correct shape and dimension according to the size and type of anchors to be fitted. In Fig. 7.1 a typical section view is illustrated and the fundamental undercut dimension identified.

There are mainly two types of undercut technologies to provide a keying type anchorage in the interior of the dimension stone or slab's thickness.

The undrilled depth, $(t - d_u)$, is normally greater than $0.4 \times t$, due to the fact that the undercut anchorage resistance is governed by the stone thickness in tension under negative wind pressure on façades [1]. For positive pressures, the majority

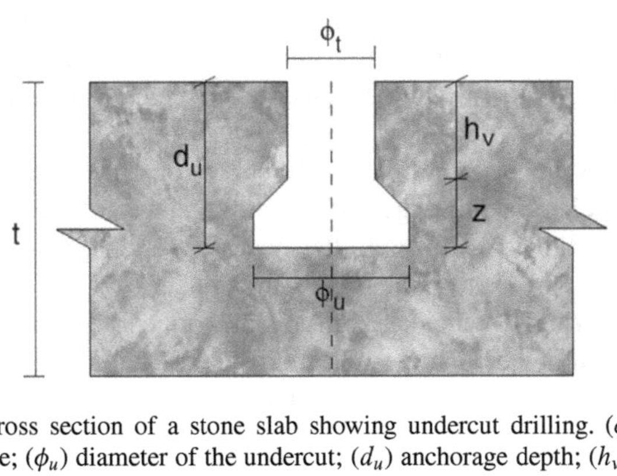


**Fig. 7.1** Typical cross section of a stone slab showing undercut drilling. $(\phi_t)$ diameter of the cylindrical drill hole; $(\phi_u)$ diameter of the undercut; $(d_u)$ anchorage depth; $(h_v)$ constant diameter hole depth; $(z)$ variable diameter hole depth; $(t)$ panel thickness

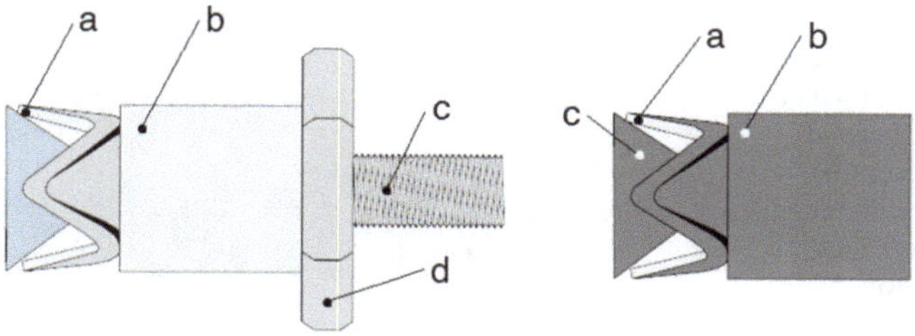


**Fig. 7.2** Expansion ring anchors with external (*left*) and internal thread (*right*); (**a**) Expansion ring; (**b**) sleeve; (**c**) cone bolt; (**d**) nut

of resultant forces is transmitted directly to a frame supporting system or any other façade backup structure.

Common values of depth sizes in the undercut hole are: $z$, around 4 mm and $h_v$, around 11 mm so that a minimum recommended slab thickness is:

$$t_{\min} \geq \frac{5}{3} \times (z + h_v) = 25 \text{ mm} \tag{7.1}$$

## 7.2.1 *Expansion Ring Keying*

An example of an expansion ring system is illustrated in Fig. 7.2. The undercut anchors have a cone bolt, either with external thread or internal thread, and generally 6 or 8 mm in diameter, an expansion ring with three or four convolutions, a sleeve and, optionally, a nut. Cone bolts and expansion rings are made of stainless steel. The sleeve is made of stainless steel or carbon. The nut is in stainless steel or aluminium.

Anchors are installed by driving the anchor sleeve against the locking ring forcing it to expand within the undercut hole form and squeezing it inside the stone. This enables a stress-free anchorage in a panel unloaded condition. This system is generally identified as the Fischer-type technology [2].

## 7.2.2 *Slotted Sleeve Keying*

The slotted sleeve keying system consists of a special anchor made of a crosswise slotted anchor sleeve with an internal thread. The anchor's upper edge has a hexagon formed to it and the respective hexagon bolt with a tooth lock washer formed to it. The anchor sleeve and the hexagon bolt with a tooth lock washer formed to it are also made of stainless steel. The anchor is fitted into an undercut drill hole and, by driving the sleeve in it is deformed (Fig 7.3).

The anchor sleeve is expanded to its original dimension by inserting the screw to a controlled depth, so that the sleeve sits snugly against the undercut section of the hole in the façade panel. This system is identified in general with the Keil-type technology [3].

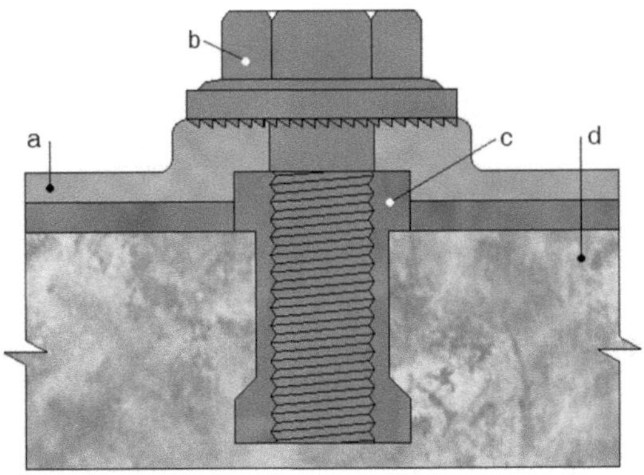


**Fig. 7.3** System using a crosswise slotted sleeve and an internal thread with a hexagon bolt. Legend: (**a**) panel bracket; (**b**) hexagon bolt with internal thread and tooth lock; (**c**) slotted sleeve; (**d**) dimension stone

### *7.2.3 Frame Supporting Systems*

In both systems, the screw or nut is fastened in until exerting slight pressure on a panel bracket forming a rigid unit with a backup supporting façade system.

The support system for a dimension stone panel in most cases consists of a framework in aluminium or stainless steel. Generally, the framework consists of four brackets attached to the back of the façade panel by means of an appropriate undercut anchor which are then placed into or onto the corresponding continuous horizontal rails (Fig. 7.4).

In most cases the horizontal rails are attached to vertical mullions or profiles, which are fixed to the building's main structure, structural concrete or masonry.

As for the other fixing systems, it must be possible to adjust undercut anchorage systems both horizontally and vertically (see Fig. 7.5). Relative movement between the panel and the framework must also be taken into account.

## 7.3 Design Procedures and Assumptions

In the following sections some issues are considered in order to define the main design aspects related with stone cladding systems using undercut type anchoring.

In effect, some particularities are related with the cladding slabs bending and others with the pull-out anchorage strength of the stone.

### *7.3.1 Deformation Limit State*

From the point of view of appearance and general utility, excessive deflections, such as in roofs or in floor coverings, are to be avoided. Thus and taking into consideration the stone creep a calculated sag of the cladding panels subjected to quasi-permanent load combination may not exceed the 1/250 of the shortest

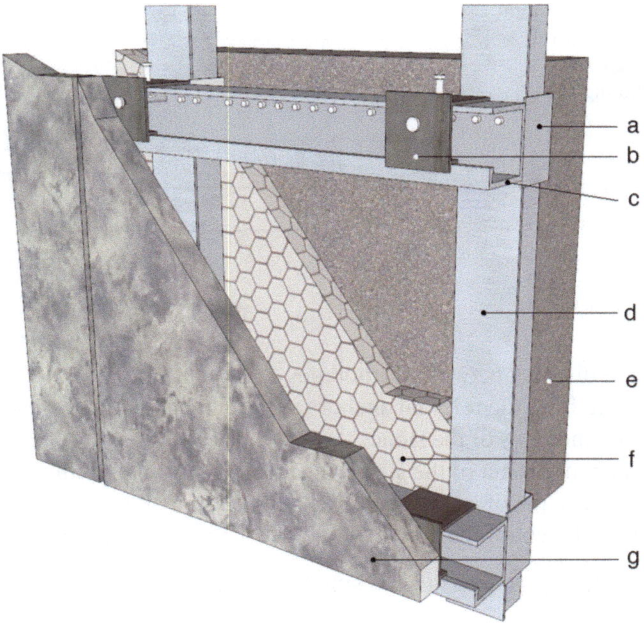


**Fig. 7.4** Example of framing supporting structure: (**a**) wall bracket; (**b**) slab's bracket; (**c**) horizontal profile; (**d**) mullion; (**e**) backup wall; (**f**) insulating layer; (**g**) dimension stone slab

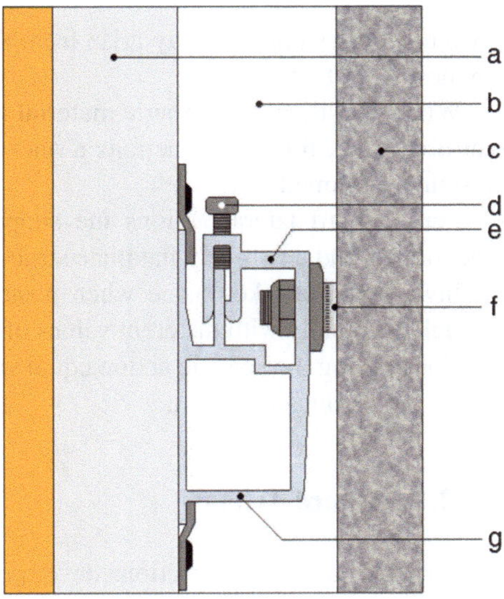


**Fig. 7.5** Framing supporting structure section view: (**a**) mullion; (**b**) air space; (**c**) dimension stone slab; (**d**) levelling screw; (**e**) slab's bracket; (**f**) undercut anchor; (**g**) horizontal profile

side's dimension, $L_{\min}$. The maximum sag, $f_{\max}$, is assessed relative to the supports using Eq. (7.2).

$$f_{\max} = \frac{1}{384} \times \frac{q_{Sd} \cdot l^2 \left(5 \times l^2 - 48 \times a^2\right)}{E \cdot I}(1 + \psi) \leq \frac{L_{\min}}{250} \tag{7.2}$$

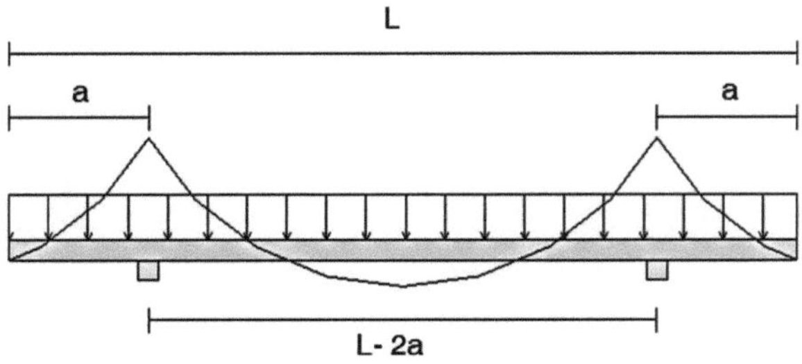


**Fig. 7.6** Idealized structural system for inward lateral actions for a slab with undercut anchoring

where:

$l$ is the greater distance between anchorages alignment;
$a$ is the distance from the relevant anchorages alignment to the nearest slab's edge;
$E$ is the elastic modulus of the stone;
$I$ is the moment of inertia of the slabs cross section per unit width;
$q_{Sd}$ is the design value of quasi-permanent load combination;
$\psi$ the creep coefficient of the type of stone which value may be taken as two [4].

The value of, $q_{Sd}$, is determined using Eq. 5.4 and Table 5.2.

## 7.3.2 *Flexural Stresses*

Two different situations may arise from the structural idealization of the undercut anchoring system.

When a resilient elastomeric material is interposed between the supporting rails and the stone, for inward actions a one-way two-side cantilevered slab structural system is assumed.

For outward lateral actions the supporting reactions are located at the four anchorages and a two-way flat plate structural system is considered.

Inward action takes place when positive dynamic wind pressure exerts, even though eventually with different values of the opposite outward wind action.

Considering the seismic action equal values are considered either for the inward or the outward lateral action.

### 7.3.2.1 Inward Actions

The inward supporting reactions develop uniformly along the two horizontal rails as illustrated in Fig. 7.6, assuming that there is full contact between the rails and the stone.

For this situation the stress calculation is straightforward assuming that the conditioning bending moment value occurs over the supports or at mid span depending on the ratio value of $a/L$:

For $a \leq 0.21 \times L$, the maximum moment, $M_{Sdc}$, is located at a cross section at midspans of the slab:

$$M_{Sdc} = w_{sd} \cdot \frac{L \cdot (L - 4 \times a)}{8} \tag{7.3}$$

Otherwise, $a > 0.21 \times L$ and the maximum moment value, $M_{Sda}$, is located in the anchorage cross section alignment:

$$M_{Sda} = w_{sd} \cdot \frac{a^2}{2} \tag{7.4}$$

The maximum stress depends on the bending moment value defined in the above equations. It is determined with the well-known formula assuming a linear uniform distribution along the slabs unsupported direction:

$$\sigma_{Sd} = \max \left\{ 6 \times \frac{M_{Sdc}}{t^2}; 6 \times \frac{M_{Sda}}{(t - d_u)^2} \right\} \tag{7.5}$$

#### 7.3.2.2 Outward Actions

On the other hand for outward actions the slab is supported in four points located at the undercut anchors axis and a two-way flat plate analogy is suitable. It must be noted that this situation occurs for inward actions when full contact between the stone and the supporting rails is not assured.

In this case bending takes place along two orthogonal directions (Fig. 7.7) with no supporting beams, i.e., profiles, and an approximated solution may be used in place of a more refined analysis such as the finite element method.

Considering that the ratio between the shorter and the longer size is greater than 0.5, a solution based on the strip method is used in order to compute the maximum stress values in the slab's stone.

As well-known the strip method is based on the fulfilment of the equilibrium requirements everywhere in a slab and their results are in the safe side [5].

Regarding Fig. 7.7, the slab is divided into strips in each of the two orthogonal directions, parallel to sides $L_i$ and $L_j$.

In a given direction, say parallel to, $L_i$, one middle strip, $m_i$, and two side strips, $s_i$, widths are defined as follows:

$$m_i = \frac{L_j}{2} - a_j; \quad s_i = \frac{L_j}{4} + \frac{a_j}{2} \tag{7.6}$$

In the opposite direction, $L_j$, following the same procedure for the middle and side respectively, one has:

$$m_j = \frac{L_i}{2} - a_i; \quad s_j = \frac{L_i}{4} + \frac{a_i}{2} \tag{7.7}$$

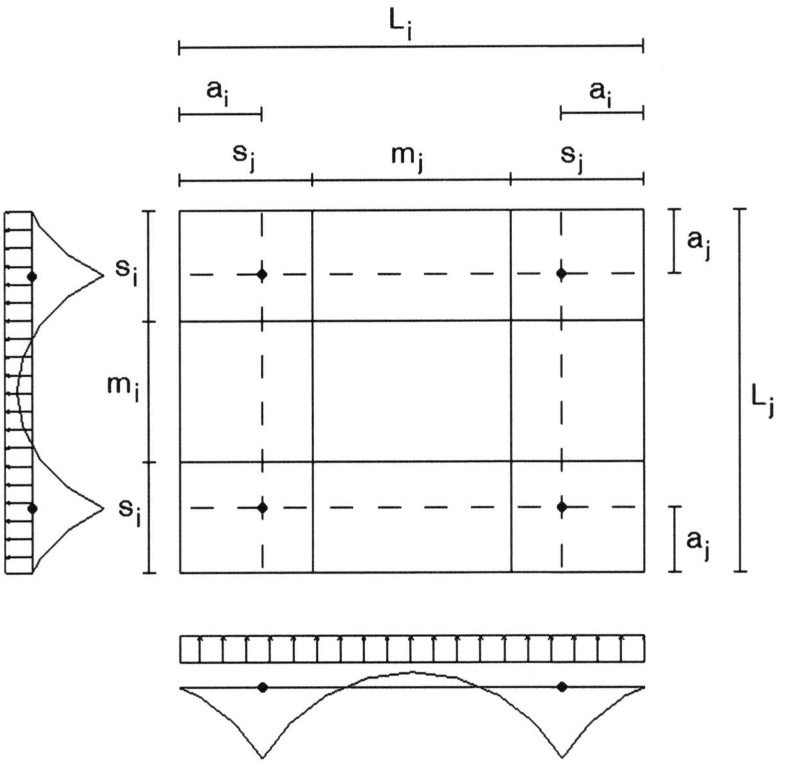


**Fig. 7.7** Idealized structural system for outward lateral actions for a slab with undercut anchoring

**Table 7.1** Distribution stress coefficients for two-way behaviour slabs

| Cross section location | Middle strips | Lateral strips |
|---|---|---|
| Mid span | $\xi_{mc} = 0.45$ | $\xi_{lc} = 0.55$ |
| Anchorage alignment | $\xi_{ma} = 0.25$ | $\xi_{la} = 0.75$ |

To compute the maximum stress values in a given direction, distribution coefficients, $\xi_m$ and $\xi_l$, are defined for middle and lateral strips depending on the cross section location. Their values are depicted in Table 7.1. Additional subscripts for these coefficients are added depending on the cross section location where the stresses are computed: ($\xi_{..c}$) for the centre of the slabs span and ($\xi_{..a}$) for the anchorage cross section alignments.

Considering the strips width and the coefficients of Table 7.1, the acting stress values for a given direction, *i*, may be determined for the lateral action value, $w_{sd}$, and the geometry of the slab.

Thus, for the middle strip at the centre of the slab the maximum stress is given by:

$$\sigma_{mc_i} = \frac{3}{2} \times \frac{\xi_{mc_i} \cdot w_{Sd} \cdot L_j \cdot L_i \cdot (L_i - 4 \times a_i)}{(L_j - 2 \times a_j) \cdot t^2} \tag{7.8}$$

and at the anchorages alignment:

$$\sigma_{ma_i} = 6 \times \frac{\xi_{ma_i} \cdot w_{Sd} \cdot L_j \cdot a_i^2}{(L_j - 2 \times a_j) \cdot (t - d_u)^2} \tag{7.9}$$

For a lateral strip at mid span:

$$\sigma_{lc_i} = \frac{3}{2} \times \frac{\xi_{lc_i} \cdot w_{Sd} \cdot L_j \cdot L_i \cdot (L_i - 4 \times a_i)}{(L_j + 2 \times a_j) \cdot t^2} \tag{7.10}$$

and at the anchors alignment:

$$\sigma_{la_i} = 6 \times \frac{\xi_{la_i} \cdot w_{Sd} \cdot L_j \cdot a_i^2}{(L_j + 2 \times a_j) \cdot (t - d_u)^2} \tag{7.11}$$

The determination of the stresses along the orthogonal direction, *j*, is found by merely swapping subscripts (*i*) with subscripts (*j*) in Eqs. (7.8), (7.9), (7.10), and (7.11).

### 7.3.3 Anchorage Pull-Out Strength

The pull-out breaking load in undercut anchorages mobilizes mainly the stone's tensile strength in the same way with other anchoring dimension stone systems.

Several studies were performed to investigate the relationship between flexural strength and breaking load at the undercut anchorage and also to gain a better understanding of the undercut anchorage's rupture behaviour. Hence, pull-out tests were also performed to study the strength behaviour of this anchorage system [1, 6].

The typical failure mode is brittle with a detaching radial cone shape spall (Fig. 7.8).

The measuring of the medium length of the spalls and their thickness allows to determine the average angle of the cone failure surface, $\alpha$, as illustrated in Fig. 7.9.

The studies allowed the definition of an average value for an angle, $\alpha$, between 15° and 20° when typical cone bolt sizes of 6 and 8 mm diameter are used. It's interesting to point out that for the same stone type the breaking load does not vary with the thread size and the spall angles are very similar for the most common used limestone, marble and granite types [1].

A minimum edge distance, $e_m$, is defined depending on the anchorage depth value, $d_u$, (Fig. 7.9), as well as a minimum space between anchors, $s_m$.

For the common anchorage depth values of 15 mm a minimum edge distance value of 55 mm and a minimum space between anchors, $s_m$, equal to 120 mm is recommended.

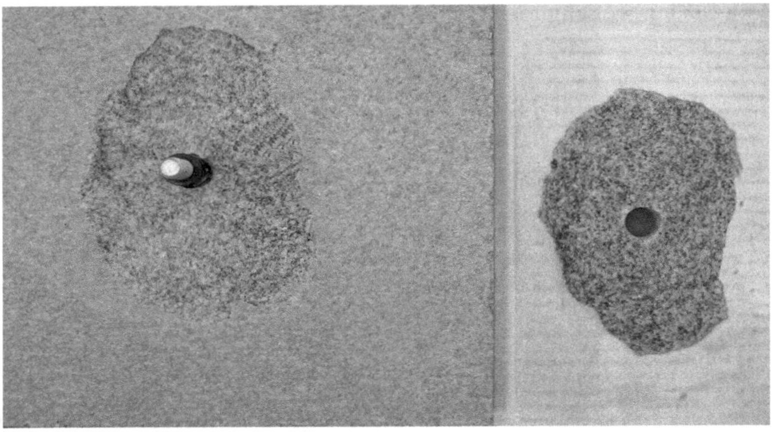

**Fig. 7.8** Typical cone mode failure in tested specimen. Tested slab and anchor (*left*) and detached spall (*right*)

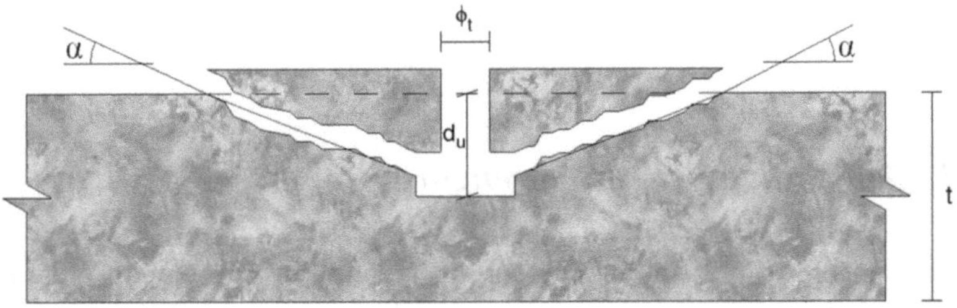


**Fig. 7.9** Typical cone mode failure in tested specimens

## *7.3.4 Finite Element Analysis*

Finite element stress analysis carried out to investigate stress distribution near the undercut and along the observed spall surface using solid elements allowing to investigate the stress distribution present in the typical undercut anchorage configuration load tests using a pull-off strength tester (Fig. 7.10).

For simplicity, the load was applied directly to the solid element nodes at the nodes in contact with the expansion ring of the cone bolt thread assuming full contact along the stone surface.

Figure 7.11 shows a 2D section cut according to the calculated stone's maximum principal tensile stresses.

Stress concentrations are present in the hole's vicinity and dissipate quickly, suggesting that anchorage failures are influenced by a non-uniform stress distribution over the failure surface. The direction of the maximum principal stresses located adjacent to the hole is approximately perpendicular to the failure surface observed in tests, meaning that failure initiates at this location, as observed.

The universal availability of powerful, effective computational capabilities, usually based on the finite element method, FEM, has altered the need for the use of

**Fig. 7.10** Digital pull-off strength tester and stiff steel plate to deter bending of the specimens

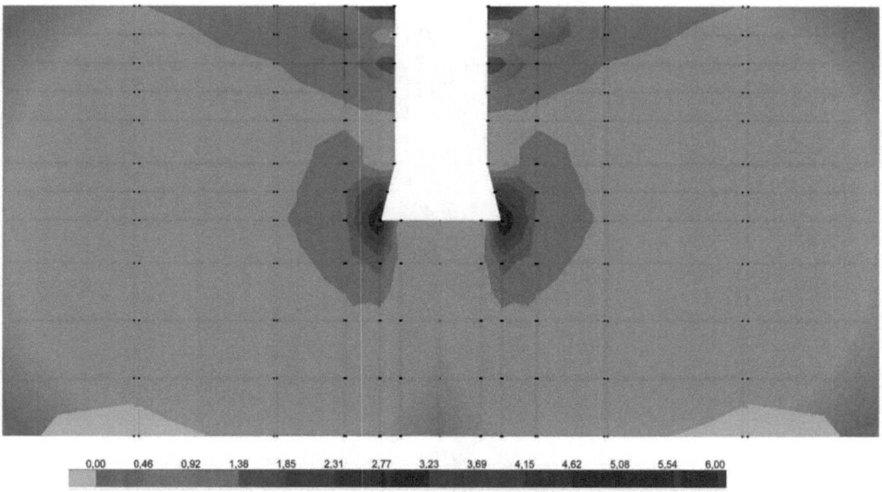


**Fig. 7.11** A 2D view stress map of a radial section cut of a drilled region after finite element calculation of maximum tensile stress; the darker zones meaning higher tensile stress

stress concentration factors. Nevertheless, it's desirable to make use of simple but effective tools to estimate the maximum stresses in the hole's vicinity thus allowing a quick and simple slab thickness estimate.

The following empirical formulation was derived based on test results validated by the FEM [1].

## 7.4 Empirical Formulation

By projecting the failure surface in a plane perpendicular to the cone axis and assuming a uniform distribution of the tensile strength over an equivalent circular idealized area according to Fig. 7.12, it is possible to establish a relation based on the maximum principal stress theory which is satisfactorily applicable to brittle or quasi-brittle materials, such as stone. Based on limiting normal stress, failure occurs when the normal stress reaches a specified upper limit and is predicted when the principal stresses equal the ultimate strength of the material.

The area of the projected cone failure surface, $A_{cf}$, is represented and may be given by:

$$A_{\mathrm{cf}} \simeq \pi\left[\left(h_{\mathrm{v}}\cdot\cot\alpha+\frac{\phi_{\mathrm{u}}}{2}\right)^{2}-\left(\frac{\phi_{\mathrm{u}}}{2}\right)^{2}\right]$$
$$=\pi\left(h_{\mathrm{v}}^{2}\cdot\cot^{2}\alpha+h_{\mathrm{v}}\cdot\phi_{\mathrm{u}}\cdot\cot\alpha\right) \qquad (7.12)$$

A stress concentration factor, $k_u$, is considered for the given geometry of the undercut drill hole. This factor is defined by the ratio between the actual maximum stress, $\sigma_{\max}$, and the medium stress, $\sigma_{tm}$, according, to Fig. 7.13.

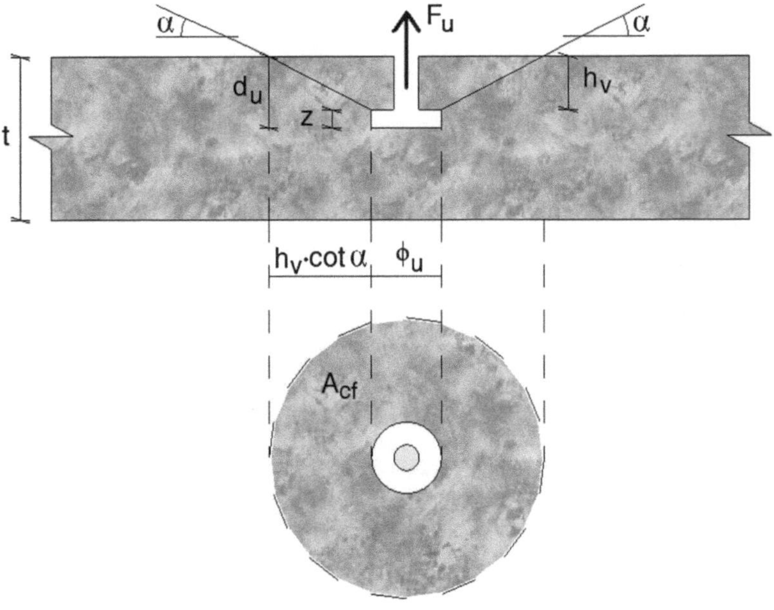


**Fig. 7.12** Spall configuration and detachment angle in slab

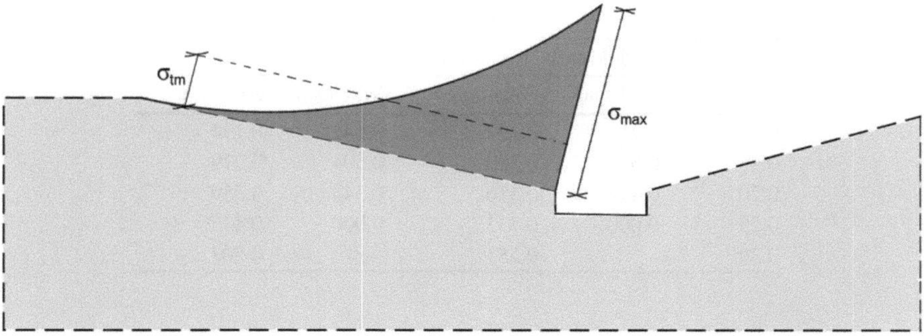


**Fig. 7.13** Stress concentration factor as the relation between the actual maximum stress, $\sigma_{\max}$ and the medium stress $\sigma_{tm}$

This way the breaking load value, $F_{br}$, may be estimated through as follows:

$$F_{\mathrm{br}} = \frac{\sigma_t}{k_{\mathrm{u}}} \cdot A_{\mathrm{cf}} \tag{7.13}$$

In Eqs. (7.12) and (7.13) the following designation is defined for the variables:

$h_v$ is the maximum spall thickness equal to $(d_{\mathrm{u}} - z)$ in Fig. 7.12;
$\phi_{\mathrm{u}}$ is the diameter of the undercut;
$A_{cf}$ is the area of the projected cone failure surface;
$\sigma_{Rt}$ is the tensile strength value of the stone;
$k_u$ is the stress concentration factor for the undercut.

Substitution of Eq. (7.12) in Eq. (7.13) allows to define the maximum tensile stress, $\sigma_{t\,\max}$, acting on the undercut region:

$$\sigma_{t\,\max} = \frac{F_{\mathrm{br}} \cdot k_{\mathrm{u}}}{\pi \cdot \left(h_{\mathrm{v}}^2 \cdot \cot^2\alpha + h_{\mathrm{v}} \cdot \phi_{\mathrm{u}} \cdot \cot\alpha\right)} \tag{7.14}$$

For current undercut diameters values, $\phi_{\mathrm{u}}$, between 13 mm and 16 mm and spall thickness, the constant diameter hole depth, $h_v$, may be expressed as a function of the slab's thickness, $t$, considering for $z$ (Fig. 7.12) a value of 4 mm:

$$h_v = \frac{3}{5} \times t - 4\ [\mathrm{mm}] \tag{7.15}$$

## 7.5 Ultimate Limit State Design

The ultimate limit state design presented is simple enough to be handled by practitioners without considerable problems and leads to conservative and robust design. It's based on studies that have shown that the anchorage zone strength is

**Table 7.2** Coefficient values of $\beta$ for use in Eq. (7.18)

| | Middle strip cross section | | Lateral strips cross section | |
|---|---|---|---|---|
| $\frac{a_i}{L_i} = \frac{a_j}{Lj}$ | Mid span | Anchorage | Mid span | Anchorage |
| 0.10 | 0.711 | 0.137 | 0.642 | 0.194 |
| 0.15 | 0.621 | 0.220 | 0.504 | 0.279 |
| 0.20 | 0.474 | 0.316 | 0.343 | 0.359 |
| 0.25 | 0.000 | 0.433 | 0.000 | 0.433 |
| 0.30 | – | 0.581 | – | 0.503 |

usually decisive in the calculation of the slab's thickness. All these studies conclude that designing stone cladding systems must take into account different effects in order to evaluate the effective stress in the critical region of the anchorage geometry.

### 7.5.1 *Flexural Design*

The limit state condition for the flexural design arises from the inequation:

$$\sigma_{Sd} \leq \sigma_{Rd} \tag{7.16}$$

where, $\sigma_{Sd}$, and, $\sigma_{Rd}$, are the design value of the acting bending stress and the design flexural strength of the stone, so that the minimum required thickness of a slab is determined accordingly.

For one-way slab behaviour the required minimum thickness is derived substituting Eq. (7.5) in Eq. (7.16), thus:

$$t \geq \max\left\{\frac{1}{2} \times \sqrt{\frac{3 \times w_{Sd} \cdot L\,(L - 4 \times a)}{\sigma_{Rd}}};\ \sqrt{\frac{3 \times w_{Sd}}{\sigma_{Rd}}} \cdot a\right\} \tag{7.17}$$

Two-way slab behaviour, takes place for outward lateral actions and the minimum slab's thickness value is dictated by the major value of Eqs. (7.8), (7.9), (7.10) and (7.11) depending on the governing direction. Bearing this in mind, the minimum slabs' thickness may be determined considering the maximum value obtained from the two sets of the four equations for each direction.

By substituting in the corresponding Eqs. (7.8), (7.9), (7.10), and (7.11) the values of Table 7.1 and considering three common relative distance values between the anchorage centres and the nearest slab's edges $\frac{a_i}{L_i} = \frac{a_j}{Lj}$ with values of 0.1, 0.15 and 0.2 [3, 7, 8], the governing stress may be located at any strip either at midspans or over the anchorages alignment, depending on the slab's size dimensions.

The minimum slab's thickness is then given by Expression (7.18), with the values of, $\beta$, presented in Table 7.2:

$$t \geq \max\left\{\beta \cdot \sqrt{\frac{w_{Sd}}{\sigma_{Rd}}} L_i;\ \ \beta \cdot \sqrt{\frac{w_{Sd}}{\sigma_{Rd}}} L_j\right\} \tag{7.18}$$

**Table 7.3** Spall angles and stress concentration factors for use in Eq. (7.20)

| Stone identification | Spall angle $\alpha$ | Concentration factor $k_u$ |
|---|---|---|
| Fine to medium grain size granite | 19° – 20° | 4.9 |
| Medium to gross grain size granite | 16° – 18° | 4.6 |
| Oolitic limestone | 19° – 20° | 6.9 |
| Calcitic marble | 15° – 16° | 9.0 |

### *7.5.2 Pull-Out Strength Design*

The design value of the action or load transmitted to a single undercut anchor may be obtained through the following equation:

$$F_{Sd} = \frac{w_{Sd} \cdot L_i \cdot L_j}{4} \tag{7.19}$$

where, $w_{Sd}$ is the design value of the uniform the lateral load acting in a dimension stone slab and, $(L_i; L_j)$, are the size lengths of the slab.

The value of the design breaking load of the anchorage, $F_{brd}$, as function of the tensile strength of the stone is expressed directly from Eqs. (7.13) and (7.14):

$$F_{brd} = \frac{\sigma_{Rtk} \cdot \pi \left(h_v^2 \cdot \cot^2\alpha + h_v \cdot \phi_u \cdot \cot\alpha\right)}{k_u} \cdot \frac{\eta}{\gamma_M} \tag{7.20}$$

with $\sigma_{Rtk}$ the design value of the tensile strength of the stone, $\gamma_M$ partial factor of safety for the stone, according to Table 4.3 and $\eta$ aging factor to take into account the loss of strength of the stone according to Table 4.4.

The values of the concentration stress factor and the spall angle are depicted in Table 7.3 for different granite types, limestone and a calcitic marble.

The safety verification format implies that $F_{brd} \geq F_{Sd}$, allowing for the determination of the minimum thickness stone since the geometry of the undercut is known.

For the common undercut geometries when Eq. (7.15) is replaced in Eq. (7.20) the following formula, for the determination of the minimum thickness, $t$, is applicable regarding that dimensions are in millimetre and stress in [MPa]:

$$t \geq \frac{5}{6} \times \frac{(\mathrm{B} - \mathrm{A})}{\pi \cdot \sigma_{Rtd}} + \frac{20}{3};$$

$$\mathrm{A} = \pi \cdot \sigma_{Rtd} \cdot \phi_u \cdot \cot\alpha;$$

$$\mathrm{B} = \sqrt{\pi \cdot \sigma_{Rdt} \left(\sigma_{Rdt} \cdot \pi \cdot \cot^2\alpha \cdot \phi_u^2 - 4 \times \sigma_{Rdt} \cdot \pi \cdot \cot^2\alpha + \omega_{Sd} \cdot L_i \cdot L_j \cdot k_u\right)} \tag{7.21}$$

with $\sigma_{Rtd}$, the design value of the tensile strength of the stone, given as usual:

$$\sigma_{Rtd} = \sigma_{Rtk} \cdot \frac{\eta}{\gamma_M} \tag{7.22}$$

and:

$\gamma_M$ – partial factor of safety for the stone, according Table 4.3;
$\eta$ – aging factor to take into account the loss of strength of the stone according to Table 4.4.

## 7.6 Application Example

The same fine grained granite presented in Chap. 5 is used as a façade cladding with undercut anchorage technology. Again, the determining action lateral action, $w_{Sd}$, has a value of 2 kPa.

The flexural strength is taken with the value of 2.5 MPa and the tensile strength with the value of 1.5 MPa respectively. To retain that these values were obtained dividing the characteristic value by a partial safety coefficient, $\gamma_M = 2.9$, according to Table 4.3 upon the calculation of the corresponding coefficient of variation of the experimental test results.

Dimension stone slabs are supported with the holes positioned at a relative distance to the edges, $a/L$, equal to15 %. The slabs orientation is such that the horizontal profile is parallel to its shortest side dimension.

A range of values between 600 and 1,500 mm for the shortest side length is considered in the example for the calculation of the required thickness of the slabs following two different criteria: based on the flexural strength by means of Eqs. (7.17) and (7.18) and based on the pull-out strength through Eq. (7.21).

A spall angle equal to 19° and a concentration factor, $k_u$, equal to 4.9 are considered according to Table 7.3. The undercut diameter value, $\phi_u$, is taken equal to 15 mm and the anchorage depth, $d_u$ is taken equal to 10 mm. For aesthetical reasons a constant ratio value between the slabs size lengths, is assumed to be equal to 2/3. The relevant data is summarized in Table 7.4.

The required minimum thicknesses were calculated and the results are presented in graphic form in Fig. 7.14.

It should be remembered that the obtained results must not be generalized, in view of the fact that they reflect a particular geometry and flexural and tensile strength for the stone.

**Table 7.4** Data used in the relevant formula for the application example

| $w_{sd}$ | $\sigma_{Rd}$ | $\sigma_{Rtd}$ | $\frac{L_{min}}{L_{maj}}$ | $\alpha$ | $\phi_u$ | $l_d$ | $k_u$ |
|---|---|---|---|---|---|---|---|
| 2.0 kPa | 2.5 MPa | 1.5 MPa | 2/3 | 19° | 15 mm | 40 mm | 4.9 |

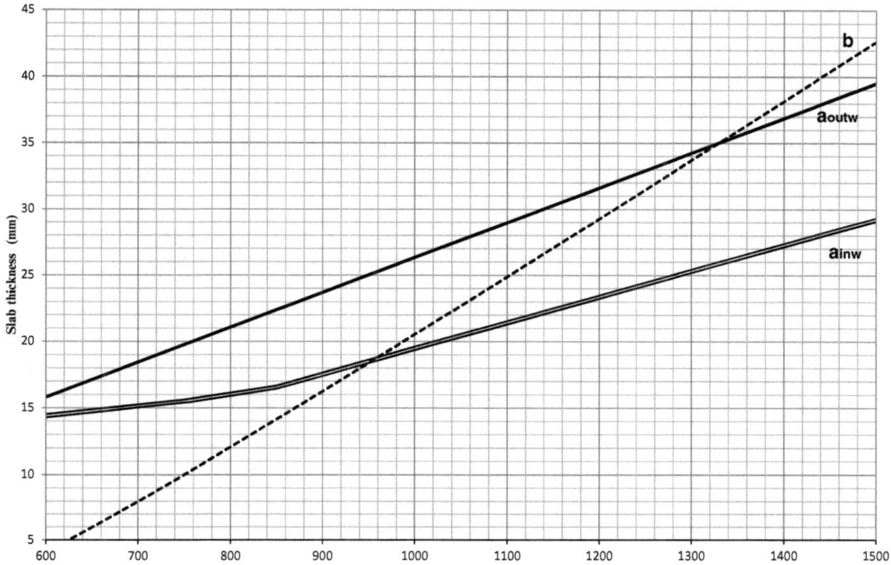


**Fig. 7.14** Required minimum thickness of dimension stone slabs considering the flexural strength, lines ($a_{inw}$) and ($a_{outw}$), for inward and outward lateral actions respectively and the pull-out strength in line (b) as a function of the minor size length according to data from Table 7.4

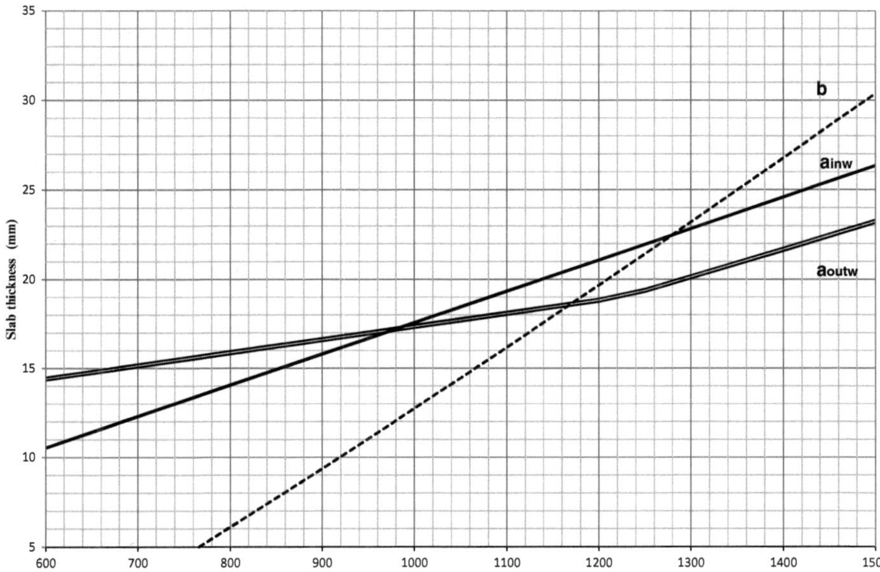


**Fig. 7.15** Required minimum thickness for square shape slabs. Lines ($a_{inw}$) and ($a_{outw}$), for inward and outward lateral actions due to flexural strength; line (b) for outward action due to pull-out strength according to data from Table 7.4

The thickness values depending on the flexural stresses are greater for the outward actions since the distance from the supports to the edges is relatively small, 15 % of the slab's side dimension, and therefore the resulting acting stresses in the anchoring zones are smaller.

The pull-out strength is the governing factor but only for slabs with minor size greater than 1,350 mm and a relative small thickness, confirming the efficiency of this anchorage system.

For square shaped slabs, and retaining the same data from Table 7.4, flexural strength is almost always determinant for the thickness estimate. Only for relevant size dimensions it's verified that the strength anchorage is prevailing. This situation is illustrated in Fig. 7.15.

The above results lead to the conclusion that undercut anchoring, despite eventually more expensive, appears to be a very efficient system.

## References

1. Camposinhos RS (2012) Undercut anchorage in dimension stone cladding. Proc ICE Constr Mater. doi:10.1680/coma.11.00050
2. Deutsches Institut für Bautechnik (2009) European Technical Approval ETA-05/0266 – Fischer-Zykon-panel anchor FZP(−W). Special Anchor for the rear fixing of façade panels made of selected natural stones according to EN 1469. fischerwerke GmbH & Co. KG, Berlin
3. Deutsches Institut für Bautechnik (2004) European Technical Approval ETA-03/0055 – Keil Hinterschnittanker KH. Special anchor for the fixing of façade plates from their back side made of dust-pressed ceramic plates (stoneware) according to EN 176. KEIL Werkzeugfabrik, Berlin
4. Singh DP (1975) A study of creep rocks. Int J Rock Mech Min Sci Geomech Abstr 12:271–276
5. Hillerborg A (1996) Strip method design handbook. E&FN Spon, an Imprint of Chapman & Hall, London
6. Lammert BT, Hoigard KR (2007) Material strength considerations in dimension stone anchorage design. J ASTM Int 4:(6). doi: JAI101062
7. Stein A (2000) Fassaden aus Natur und Betonwerkstein: Konstruktion und Bemessung nach DIN 18516. Verlag Georg D.W. Callwey, München
8. Halfen (2012) Halfen natural stone support systems – façade. FS 11-E. Halfen

# Chapter 8
# Kerf Anchorage

**Abstract** Installation procedures for dimension stone kerf anchorage are presented together with design procedures and core parameters affecting the anchorage strength.

The different issues related with either continuous or discrete-length supporting profiles are treated in a consistent manner to ascertain minimum slabs thickness formulae depending on the flexural strength capacity or the anchorage strength capacity.

A special consideration is taken to the stress field induced in the support region through stress concentration factors calibrated by experimental studies and the finite element method.

An application example to determine minimum thickness is presented for common and several slabs dimensions and results are discussed.

## 8.1 Introduction

Inadequate initial evaluation of material durability and panel strength resulted in varying degrees of distress in some claddings. Despite the fact that early stone cladding was installed with little thought to structural analysis, and knowing that damage during handling operations can result in panel cracking, some of which may not become promptly, most failures in stone wall systems can be attributed to the fastening system and, as already mentioned in previous chapters, thickness slabs' dimensioning is mainly dictated by a more rigorous analysis at the supporting region [1, 2].

The traditional "bending at midspan" stress calculations, often referred to as allowable stress design in kerf supporting system may seem the only way to "solve the problem", perhaps because the supporting shape itself drives the designer's mind to a one way slab like type structure [3].

R. de Sousa Camposinhos, *Stone Cladding Engineering*, 
DOI 10.1007/978-94-007-6848-2_8, 

Recent studies of the kerf anchorage behaviour with the basis on results of a series of tests carried out according to the ASTM C1354 [4] followed by stress analysis finite element method calculations were the basis for a simple semi-empirical formula to estimate the breaking load which is adopted [2, 3].

## 8.2 Kerf Anchoring

Typical kerf anchor constitution is normally in formed stainless steel or extruded aluminium profiles. They may be continuous or discontinuous and are typically located on the top and bottom of slabs' edges for easier access and alignment during installation (Fig. 8.1).

These profiles are fastened to a support frame or connected directly to the building structure by bolts or anchors, thus providing the essential mechanical connection between the stone and the structure.

In most cases these horizontal profiles are attached to vertical mullions which are fixed to the building's structure or a backup wall (Fig. 8.2).

As for the other fixing systems, it's both horizontally and vertically adjustable allowing for relative movement between the panel and the framework.

The main issues regarding assembling stone panels using kerf anchors revolve around considerations which are implicit in the design [5].

Regarding the drawing in Fig. 8.3 due attention should be given to ensuring the placing a plastic or metal adjustment bearing shim slightly larger than the anchor's fastened face (a); preventing the connection from slipping after vertical adjustment with a diagonally slotted washer plate, welded washer or serrated anchor and washer (b); attaching the anchor with a fastener to a backup structure (c).

It's also important to maintain clearance to avoid contact between the kerf anchor (d) and the stone due to weight transfer and allowing for movement, creep, expansion, and fabrication and installation tolerances (e); to place a plastic or stiff rubber bearing shim to level, separate or to prevent bearing of stone kerf fin on

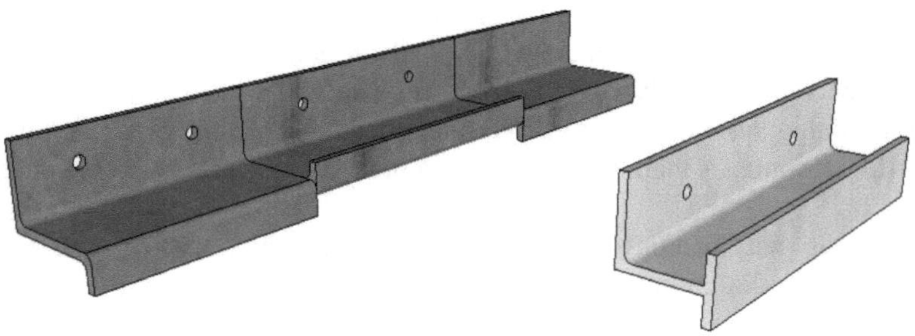

a b

**Fig. 8.1** Kerf anchor profiles in discrete (**a**) and continuous (**b**) length configurations

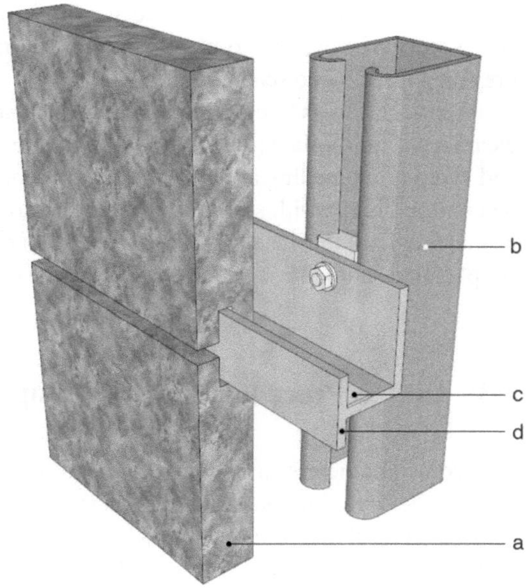


**Fig. 8.2** Kerf anchorage 3D view; (**a**) kerf edge stone; (**b**) mullion; (**c**) resilient material strip; (**d**) horizontal profile

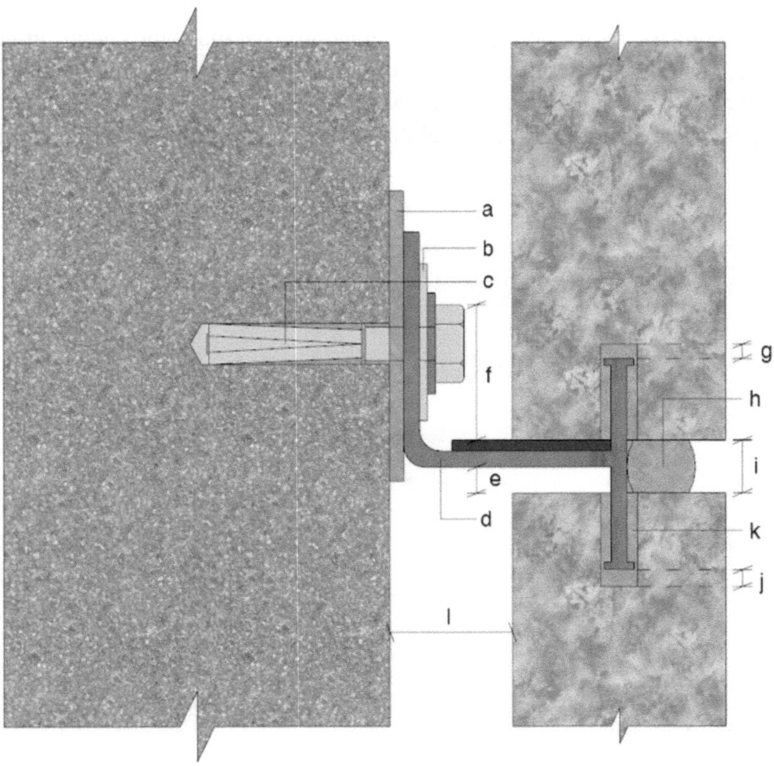


**Fig. 8.3** Typical cross-section of kerfed stone cladding systems

anchor radius (f); To keep clearance avoiding point loading at the kerf at the top (g) and bottom of stone (j); to place a backer rod or foam tape at proper depth (h) to prevent a three-sided sealant bond if used.

Important too is the sizing of the joints to allow for hardware, tolerances, clearances, appropriate movement and joint filler capability thus ensuring clearance and avoid point loading at the top and bottom of the stone (i). Also kerf filling must be continuous (k) with sealant to top of kerf to prevent moisture accumulation. At last but not least the distance (l) between the slab backing face to the anchor reacting support line should be reduced to mitigate the eccentric weight on anchor.

## 8.3 Design Procedures and Assumptions

The core parameters affecting the anchorage capacity are depicted in Fig. 8.4, namely the kerf slot width, the thickness of the stone fin, the depth of contact and the engagement length of the clip's leg [6–8].

Regarding the kerf slot width, $t_k$, it must be noted that the distance across the saw cut varies due to the original saw blade thickness and the blade's reliability of plane and rotation.

The thickness of the stone fin, $t_f$, remaining on the panel edge, after kerf anchorage assembly determines its potential strength. The distance from the kerf's inside face to the finish face varies according to the stone panel gauge after having been sawn and the stone panel thickness, in accordance with the previously mentioned factors.

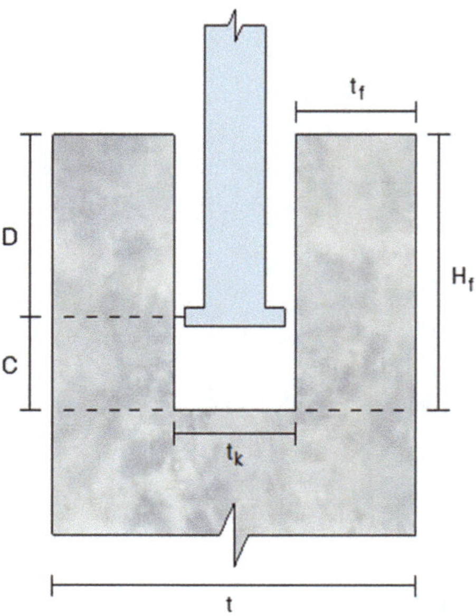


**Fig. 8.4** Kerf slot width, $t_k$, fin thickness, $t_f$, depth of rail contact, $D$, and, $C$, clearance distance

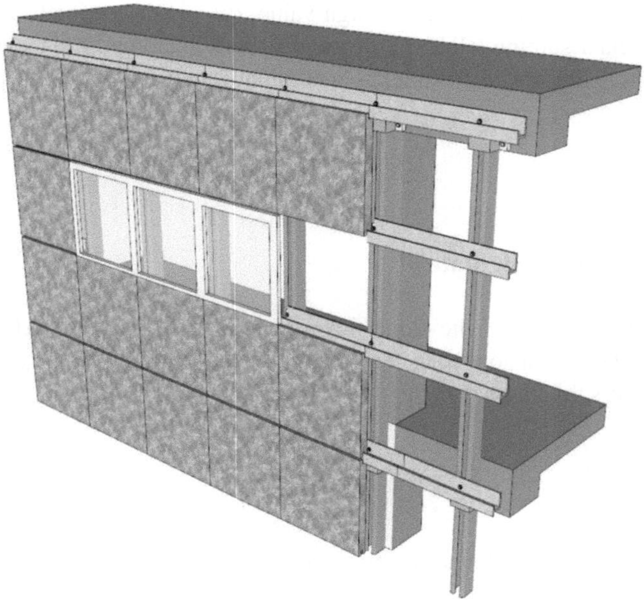

**Fig. 8.5** Schematic view of kerfed slabs mounted in a continuous rail framing system

The depth of contact, $D$, depends on the clearance distance , $C$, and on the depth of the saw cut, $H_f$.

The clearance distance, $C$, is defined as small as possible as a function of the backup structural deformation to avoid contact between the rail and the stone.

### *8.3.1 Flexural Design*

Given that the saw cut is located in the central third of the stone thickness no differences between inward or outward action need to be made on the calculation of the flexural strength.

For both conditions supporting reactions develop uniformly along two horizontal parallel rails as illustrated in Fig. 8.5.

Thus, the acting bending moment design value, $M_{Sd}$, at a midspan cross-section of a slab is expressed as follows:

$$M_{Sd} = \frac{q_{Sd} \cdot L^2}{8} \tag{8.1}$$

where, $q_{Sd}$, is the design value of the transverse acting load relative to the plane of the slab, and, $L$, is the distance between centre of the supports alignment.

The resulting maximum stress design value, $\sigma_{Sd}$, is determined with the well-known formula assuming a linear uniform distribution along the slabs unsupported direction where, *t*, is the slab's thickness nominal value:

$$\sigma_{Sd} = \frac{3 \cdot q_{Sd} \cdot L^2}{4 \cdot t^2} \tag{8.2}$$

#### 8.3.1.1 Ultimate Limit State

After the design value of the stone flexure strength, $\sigma_{Rd}$, being determined, the thickness of the dimension stone can be estimated. Assuming that, $\sigma_{Sd} \leq \sigma_{Rd}$, the minimum slab's thickness is defined as follows:

$$t \geq L \cdot \sqrt{\frac{3}{4} \times \frac{q_{Sd}}{\sigma_{Rd}}} \tag{8.3}$$

It must be emphasized that the value of , $q_{Sd}$, in Eq. (8.3) is defined as a result of the fundamental combination for ultimate limit states with one variable action in general, *e.g.*, wind or seismic action.

#### 8.3.1.2 Deformation Limit State

The same issues concerning a two way simple support slab under a uniform loading may be established to the verification of the deformation limit sate.

Thus, the calculated sag, $f_{\max}$, of the cladding panels subjected to quasi-permanent loads may not exceed the 1/250 of the span length.

The well-known formula presented in Expression (8.4) may be applied with the remarking difference that the design value of the applied load, $q_{Sd}$, is a result of a different combination rule which normally concerns the self weight of the slabs and, eventually, a fraction of the applied vertical loads:

$$f_{\max} = \frac{5 \times q_{Sd} \cdot L}{384 \times E \cdot I}(1 + \psi) \leq \frac{L}{250} \tag{8.4}$$

where, *L*, is the length of the span, *E*, the elastic modulus of the stone, *I*, is the moment of inertia of the slabs cross section per unit width, $q_{Sd}$, is the design value of the quasi-permanent load combination acting on the slab (see Table 5.2) and , $\psi$, the creep coefficient of the type of stone which value may be taken with a value of two [9].

**Fig. 8.6** Test arrangement before continuous knife edge adjustment in a marble specimen

### *8.3.2 Kerf Anchorage Strength*

Designing stone cladding slabs with kerfed edges must take into account different effects in order to evaluate the effective stress in the critical region of the kerf geometry.

The structural capacity of this type of anchorage depends essentially on the combined shear and flexural strength of the stone's fin or leg.

Several studies were performed to investigate the relationship between flexural strength and breaking load at the kerf anchorage as well as to gain a better understanding of the anchorage's rupture behaviour, either using continuous or isolated engage lengths. Hence, pull-out tests were also performed to study the strength behaviour of this anchorage system [7, 10, 11].

The stone types used in the tests were two limestone, two granites and one marble. All anchorage strength tests were performed according to the ASTM C 1354 test method [4] using specimens with both faces sawn.

The kerf anchor was installed in only one of the stone specimen's edges; the opposite edge was placed over a roller support. The load was applied at a close distance from the kerf anchorage equal to $t + H_f$, (see Fig. 8.4) to avoid interference in the failure mode. The actual load applied to the anchor, which was less than the total applied load, was calculated using simple statics principles. A test configuration is illustrated in Fig. 8.6.

It has to be pointed out that the length to which the kerf clip's leg is engaged, taking into account that the actual length is not necessarily the effective length of engagement.

In fact, when the support rail's stiffness (which depends on the rail itself, on its cross-section's moment of inertia and modulus of elasticity) is less than the

**Fig. 8.7** Slab's edge after tested with a strap anchor

dimension stone's stiffness (determined by the stone's thickness and modulus of elasticity) along the span of the dimension stone, the entire kerf rail length will not act as a continuous support. In general the continuous kerf rail is made of aluminium and does not provide a continuous support for the stone unless the kerf rail is much stiffer than the stone panel [2, 8].

Thus, two different support conditions have to be assumed, both depending on the effective contact width between the anchorage device and the stone.

When the stiffness of the slab's fin is superior than the rail blade in contact with the stone, an effective contact width, $B_{\text{eff}}$, is assumed at each of the quarters of the slab's ends.

On the other hand, the full width of the slab is mobilized in the reacting to the applied forces.

This is the same to say that a contact parameter, $\delta$, may be define in order to determine the effective contact width in a continuous supporting system:

$$\delta = \frac{B_{\text{eff}}}{B} \tag{8.5}$$

Based upon the studies of Lammert and Hoigard [2] the connection is supposed to act in all the full width when the deformation of the supporting blade rail is less than 0.1 % of the slab's width and in this case the contact parameter, $\delta$, takes a unit value, otherwise a value of 0.5 may be assumed.

When the anchorage configuration has a discrete-length, i.e., non-contiguous the effective width is defined taking into consideration the strap anchor length and the post break format of slab's edge (see Fig. 8.7).

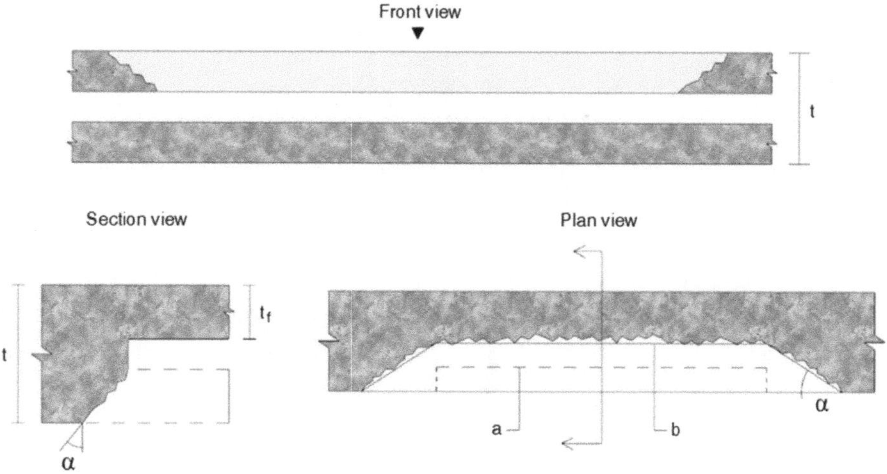


**Fig. 8.8** Idealized geometry of a detached edge in a discrete anchor. (**a**) strap anchor tail; (**b**) idealized effective contact width path

**Table 8.1** Spall angle values to be used in Eq. (8.6)

| Stone type | $\alpha$ |
|---|---|
| Fine to medium grained granite | 35°–40° |
| Oolitic limestone | 13°–20° |
| Crystalline marble | 40°–45° |
| Semi-crystalline limestone | 17°–35° |

Considering the schematic views illustrated in Fig. 8.8 the effective contact width may be expressed as follows:

$$B_{\text{eff}} = L_{\text{anc}} + 2 \times d_{\text{eff}} \cdot \cot \alpha \tag{8.6}$$

with:

$L_{\text{anc}}$ the strap anchor length;
$\alpha$ the spall detachment angle;
$d_{\text{eff}}$ the effective kerf's depth defined by $H_f - t_f \tan \alpha$;
$H_f$ and, $t_f$, have been already defined (see Fig. 8.4).

To note that the same value of the relative contact parameter, $\delta$, from Eq. (8.5) stands for the effective contact width in a non-continuous supporting system.

The ranges of the observed values of the angle of detachment $\alpha$ are given in Table 8.1 for the studied stone types.

#### 8.3.2.1 Stress Analysis

A considerable number of theories have been proposed relating uniaxial to biaxial or triaxial stress systems. For brittle or quasi-brittle materials, the maximum stress

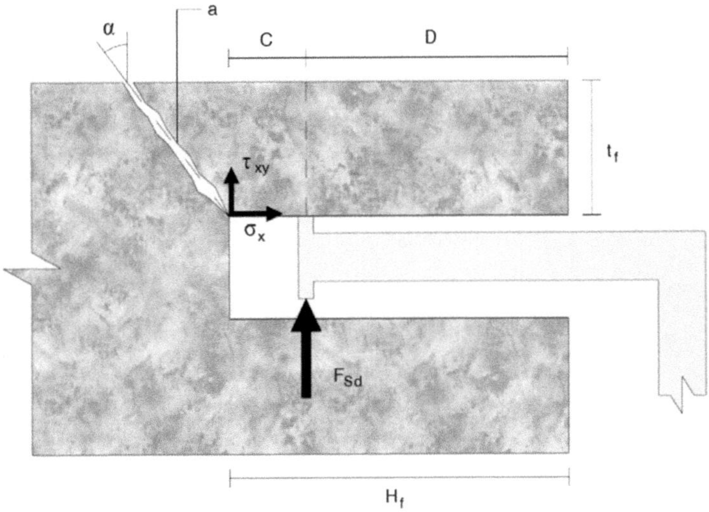


**Fig. 8.9** Normal and shear stress at a critical kerf leg section

criterion, or Rankine criterion, has been applied. In this criterion, it's assumed that failure occurs in a multiaxial state of stress when either a principal stress reaches the uniaxial tensile strength or a principal compressive stress reaches the uniaxial compressive strength. As for natural stone, the compressive strength is considerably greater than the tensile strength and in the present case the principal compressive stress is thus neglected.

Regarding Fig. 8.9 the principal tensile stress, $\sigma_{\text{prin}}$, in the critical plane of the kerf's leg is given by:

$$\sigma_{\text{prin}} = \frac{\sigma_x + \sigma_y}{2} + \frac{1}{2}\sqrt{\left(\sigma_x - \sigma_y\right)^2 + 4 \times \tau_{xy}^2} \tag{8.7}$$

with

$$\sigma_x = \frac{6 \times F_{Sd} \cdot C}{B_{\text{eff}} \cdot t_f^2}; \qquad \sigma_y = 0 \tag{8.8}$$

and

$$\tau_{xy} \sim \frac{F_{Sd}}{B_{\text{eff}} \cdot t_f} \tag{8.9}$$

Note that the shear stress in this region is assumed to have an average value when compared with the exact formula.

Substituting Eqs. (8.9) and (8.8) in Eq. (8.7) for the maximum tensile stress, $\sigma_{\text{prin}}$, one has:

$$\sigma_{\text{prin}} = 3 \times \frac{F_{Sd} \cdot C}{B_{\text{eff}} \cdot t_f^2} + \sqrt{9 \times \left(\frac{F_{Sd} \cdot C}{B_{\text{eff}} \cdot t_f^2}\right)^2 + \left(\frac{F_{Sd}}{B_{\text{eff}} \cdot t_f}\right)} \tag{8.10}$$

It should be noted that the strength of the stone is given by the effective cross section of the kerf's leg and the transverse action is determined taking into consideration its design value, $w_{Sd}$, and the span, $L$, and the full width $B$ of the slab.

Thus, for each side of the supported slab, to the value of the acting uniform load in the slab corresponds the following reaction at the kerfs region:

$$F_{Sd} = \frac{w_{Sd} \cdot L \cdot B}{2} \tag{8.11}$$

When comparing $\sigma_{\text{prin}}$ with the flexural strength, $\sigma_{\text{Rd}}$ for each type of stone, it's necessary to attend the presence of the variations in the geometry, such as shoulders, grooves, holes, etc.

This variation in geometry modifies stress distributions, which are obtained through elementary stress design formulas. To remember that a system of forces acting on a small portion of the surface of an elastic body is replaced by another statically equivalent system of forces acting on the same portion of the surface, the redistribution of loading produces substantial changes in the stresses in the immediate neighbourhood of the loading [12].

Equation (8.10) is based on the members having a constant or a section with a gradual contour change. In the present case localized high stresses occur and may be measured by a stress concentration factor, hence a 'theoretical' or 'geometric' stress concentration factor, $K_t$, is defined as the ratio between the actual maximum stress and the nominal stress:

$$K_t = \frac{\sigma_{\max}}{\sigma_{\text{prin}}} \tag{8.12}$$

where, $\sigma_{\max}$, is the maximum stress to be expected in the member at the critical section and, $\sigma_{\text{prin}}$, is the reference or nominal principal tensile stress at the critical section according to Eq. (8.10).

In some cases, a theoretical factor can be derived for, $K_t$, based on the theory of elasticity, or it may be obtained through a laboratory stress analysis experiment. The universal availability of powerful, effective computational capabilities, usually based on the finite element method, F.E.M., has altered the need for and use of stress concentration factors. Nevertheless, it's desirable to compare the values obtained through both methods.

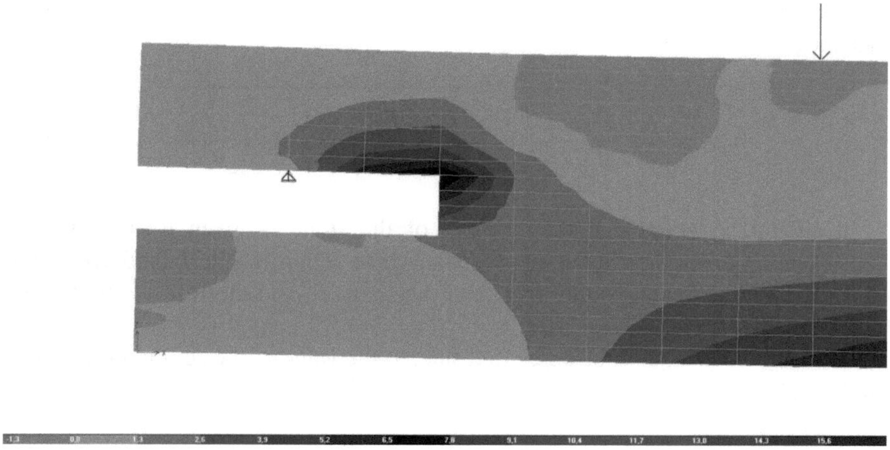

**Fig. 8.10** FEM calculation of maximum principal stresses at bottom of kerf showing stress concentration with the darker zones meaning higher tensile stress

Based on the photo-elastic tests of Leven and Hartman (1951) and of Wilson and White (1973), cited by Pilkey [13], we may find out the corresponding $K_t$ for the present case. A value of $K_t \approx 1.58$ can be obtained using chart 3.9 in the reference bibliography.

Alternatively, a two-dimensional finite element model representing a kerf anchorage configuration was developed to determine the stone's stress state in the vicinity of the anchor.

The model uses a four-node formulation for membrane elements with an isoparametric formulation for the translational in-plane stiffness component.

A linear elastic analysis was performed using a refined mesh in the anchor's vicinity in order to capture the elastic state of stress at the radius fillet as shown in Fig. 8.10.

Distinct runs were made taking into account the self-weight of the panels and the average breaking load values observed in the tests for each type of stone [7, 10]. A very good concurrence of the values was obtained when comparing the theoretical factor of $K_t \approx 1.58$. In fact, the value of the ratio between the maximum stress, $\sigma_{\max}$, obtained from F.E.M., and $\sigma_{\text{prin}}$ from Eq. (8.10) is approximately equal to 1.67.

### *8.3.3 Effective Stress Concentration Factor*

Standard design methods for engineering structures and components under static loading are usually based on avoiding failure caused by yielding/plastic collapse or buckling. The loading resistance is determined based on conventional solid mechanics theories of stress analysis. Conventional design procedures to prevent fatigue failure are based on experimental results for particular geometric details and materials. None of these procedures are capable of taking into account the effects of severe stress concentrations or crack-like flaws. The presence of such flaws is more or less inevitable to some extent in practical manufacturing and is a characteristic of quasi-brittle materials such as natural stones.

The strength of a material can, in the simplest terms, be viewed as the maximum stress which the material can sustain under given conditions. Theoretical strength calculations often overestimate the strength if they do not incorporate mechanisms to account for material defects such as cracks. To propagate material defects, the theoretical strength must be overcome only locally, within a "stress concentration" produced by these flaws; hence the effective strength of the material is lowered.

Fracture mechanics methods are particularly useful in making fitness for purpose assessments of the effects of flaws. An important effect arises when the crack affects the net cross-section area either in the case of a through-thickness crack in a plate of finite width, or in the case of remaining ligaments between the crack front and a free surface for part-thickness cracks [14–16].

When a crack is located in a region of geometric stress concentration, there will be a further increase in the stress. The overall effect depends on the relative size of the crack and the stress concentration zone. In this situation, a stress intensity magnification factor, $M_c$, is required to represent the necessary amplification [16, 17].

Based on the studies developed by the author an effective stress concentration factor, $K_{\text{eff}}$, is obtained taking into consideration not only the geometry of the kerfed section but also the magnification factor [7, 8, 10]. Hence the following expression is used to ascertain the effective stress concentration factor.

$$K_{\text{eff}} = K_t \cdot M_c \approx 1.6 \times M_c \tag{8.13}$$

Values of the magnification factor, $M_c$, may be estimated as a function of the spall angle, $\alpha$, (see Figs. 8.8 and 8.9) in each stone type according to the following expression [8]:

$$M_c = 1 + \tan\alpha \tag{8.14}$$

The effective stress concentration factor may be defined as follows:

$$K_{\text{eff}} \approx 1.6 \times (1 + \tan\alpha) \tag{8.15}$$

Based on the above expression the following values for the effective stress concentration factor have been proposed by the author and are depicted in Table 8.2.

**Table 8.2** Effective stress concentration factor

| Stone type | $K_{\text{eff}}$ |
|---|---|
| Fine to medium grained granite | 2.9 |
| Oolitic limestone | 2.1 |
| Crystalline marble | 3.2 |
| Semi-crystalline limestone | 2.4 |

### *8.3.4 Anchorage Design Formulae*

Considering the stress concentration effect trough the factor, $K_{\text{eff}}$, and substituting Expression (8.11) for the design acting load, Expression (8.5) for the contact support parameter, $\delta$, in Eq. (8.10) that stands for the nominal tensile stress, $\sigma_{\text{prin}}$, the following expression for the value of the maximum acting tensile stress in the kerf region is obtained:

$$\sigma_{Sd} = \frac{3 \times w_{Sd} \cdot L \cdot C + \sqrt{w_{Sd}{}^2 \cdot L^2 (9 \times C^2 + t_f^2)}}{2 \times \delta \cdot t_f^2} \times K_{\text{eff}} \tag{8.16}$$

With Eq. (8.16) and the safety verification format being established for the ultimate limit state at stress level:

$$\sigma_{Rd} \geq \sigma_{Sd} \tag{8.17}$$

it's possible to establish the minimum thickness of the stone, $t$, depending on the kerf slot width, $t_k$, and on the fin thickness, $t_f$:

$$t = 2 \times t_f + t_k;$$

$$t_f \geq \frac{1}{2} \times \frac{\sqrt{w_{Sd} \cdot L \cdot K_{\text{eff}} (w_{Sd} \cdot L \cdot K_{\text{eff}} + 12 \times \sigma_{Rd} \cdot \delta \cdot C)}}{\sigma_{Rd} \cdot \delta} \tag{8.18}$$

and at the same time the maximum value for the clearance, $C$, (see Fig. 8.4) in the same manner, as a function of the fin thickness, $t_f$:

$$C \leq \frac{1}{12} \times \frac{\left(2 \times \sigma_{Rd} \cdot \delta \cdot t_f\right)^2 - (w_{Sd} \cdot L \cdot K_{\text{eff}})^2}{\sigma_{Rd} \cdot \delta \cdot K_{\text{eff}} \cdot w_{Sd} \cdot L} \tag{8.19}$$

It is to be noted that in some situations where structural movements are negligible or null, for example in interior horizontal spaces the value of , $C$, may be considered with a value of zero, meaning that the state of stress is only due to shear. Hence the value of, $t_f$, in Eq. (8.18) is given by:

$$t_f \geq \frac{1}{2} \times \frac{w_{Sd} \cdot L \cdot K_{\text{eff}}}{\sigma_{Rd} \cdot \delta} \tag{8.20}$$

A résumé of the used notation in the Equations is listed below:

$L$ – distance between centre of the supports;
$B$ – slab's width;
$B_{\text{eff}}$ – effective contact width of the slab;
$\delta = \frac{B_{\text{eff}}}{B}$, support contact parameter;
$t$ – thickness of the dimension stone;
$C$ – kerf clearance distance to the blade's rail;
$t_f$ – fin's edge stone thickness;
$t_k$ – kerf slot width;
$\alpha$ – spall detachment angle;
$K_{\text{eff}}$ – effective stress concentration factor;
$\sigma_{Rd}$ – design value of the stone flexure strength defined by $\sigma_{Rk}/\gamma_M \cdot \eta$;
$\sigma_{Rk}$ – characteristic value of the stone flexure strength;
$\gamma_M$ – partial factor of safety for the stone, according to Table 4.3;
$\eta$ – aging factor taking into account the loss of strength of the stone according to Table 4.4.

## 8.4 Application Example

The same fine grained granite presented in Chaps. 5 and 6 is used as a façade cladding using kerfed dimension stone slabs. The design value of the lateral action, $w_{Sd}$, is equal to 2 kPa.

The flexural strength design value of the stone is equal to 2.5 MPa as a result of the characteristic strength and a partial safety coefficient, $\gamma_M = 2.9$, according to Table 4.3 upon the calculation of the corresponding coefficient of variation of the experimental test results.

A range of values between 600 and 1,600 mm for the slabs span is considered. The calculation of the required thickness for the slabs follows two different criteria: based on the flexural strength by means of Eq. (8.3) or considering the anchorage strength using expressions (8.18).

A stress concentration factor, $K_{\text{eff}}$, equal to 2.9 is considered according to Table 8.2.

For the kerf slot width, $t_k$, a value of 8 mm is assumed. The slabs are mounted in a continuous profile thereby the contact parameter, $\delta$, takes the value of 0.5.

Two different values for the clearance, $C$, in the kerf at the top and bottom of stone are considered with the values of 5 and 10 mm.

The relevant data is summarized in Table 8.3.

**Table 8.3** Data used in the relevant formula for the application example

| $w_{sd}$ (kPa) | $\sigma_{Rd}$ (MPa) | $t_k$ (mm) | $\delta = \frac{B}{B_{\text{eff}}}$ | $K_{\text{eff}}$ |
|---|---|---|---|---|
| 2.0 | 2.5 | 8 | 0.5 | 2.9 |

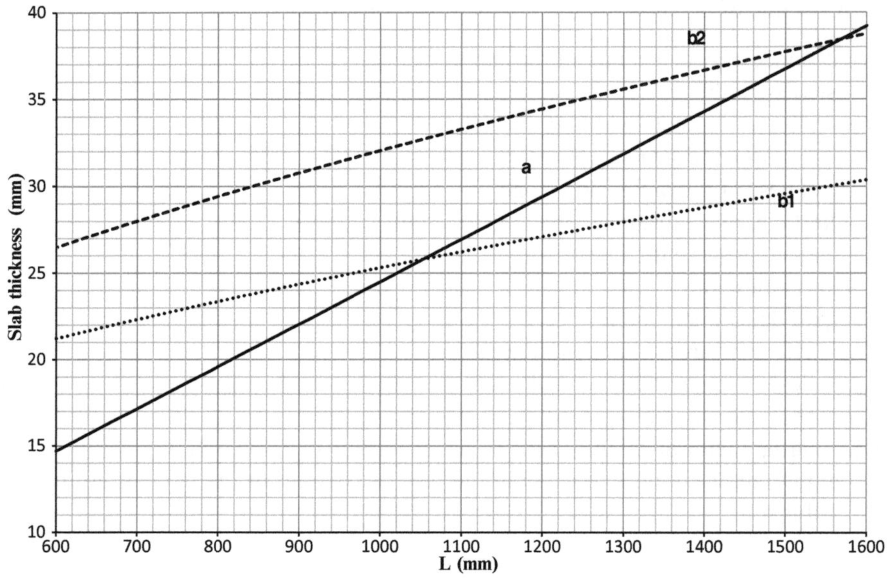


**Fig. 8.11** Required minimum thickness of dimension stone slabs considering the flexural strength, (line a) and the anchorage strength (lines $b_1$ for $C = 5$ mm and $b_2$ for $C = 10$ mm) according to data from Table 8.3

The required minimum thicknesses were calculated and the results are presented in graphic form in Fig. 8.11.

From the graph lines in Fig. 8.11 it's quite evident that only for larger slabs with narrow clearances the thickness is dictated by the flexure strength at midspans, even though and furthermore it will be difficult to accept in practice this antagonistic condition.

The slabs thickness values depend mostly on the kerfs fins strength acting as small cantilevers so that the greater the span the bigger the required thickness.

Depending on several conditions such as cost and time production a solution to respond to the required thickness at the supports is the use of pieces of stone or metal attached to the slab thus creating a slot mobilizing all the slabs thickness.

These liners have a reglet cut into them to form the kerf thus receiving the anchor blade allowing the transfer of the loads from the stone to the anchor (see Fig. 8.12).

Care must be taken in the connections between the liners and stone. They are generally pre-cut in determined lengths and attached to the stone using stainless steel fasteners and adhesive bond [18].

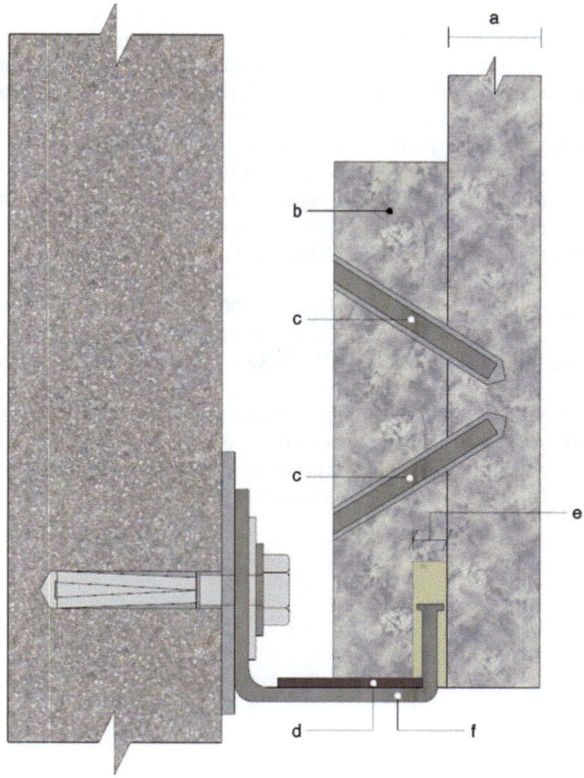


**Fig. 8.12** Stone liner as reinforcing anchoring element. (**a**) dimension stone slab; (**b**) stone liner; (**c**) dowels embedded in epoxy; (**d**) rubber shim; (**e**) reglet cut width; (**f**) anchor rail

# References

1. Lewis MD (2007) Characteristics that affect the integrity of existing thin stone cladding. J ASTM Int 4(6):1–7
2. Lammert BT, Hoigard KR (2007) Material strength considerations in dimension stone anchorage design. J ASTM Int 4(6):1–19. doi:JAI101062
3. Camposinhos RS, Camposinhos RPA (2012) Dimension stone design – kerf anchorage in limestone and marble. Proc ICE Constr Mater 165(3):161–175. doi:101.1680/coma.2009.162.3.95
4. Testing ASf, Materials (2004) C1354-06: standard test method for strength of individual stone anchorages in dimension stone. ASTM
5. Materials A-ASfTa (2010) Standard guide for selection, design, and installation of dimension stone attachment systems, vol ASTM C1242 – 10. American Standard of Testing Materials, West Conshohocken
6. Lewis MD (1995) Modern stone cladding: design and installation of exterior dimension stone systems, vol MNL 21, ASTM manual series. ASTM, Philadelphia
7. Camposinhos RS, Camposinhos RPA (2012) Dimension stone design – kerf anchorage in limestone and marble. Proc ICE Constr Mater 165(3):161–175. doi:10.1680/coma.8.00052
8. Camposinhos RS (2009) Revestimentos em Pedra natural com Fixação Mecânica – Dimensionamento e Projecto, vol 1, 1st edn. Edições Sílabo, Lisboa
9. Singh DP (1975) A study of creep of rocks. Int J Rock Mech Min Sci Geomech 12(9):271–276. doi:10.1016/0148-9062(75)91084-0

10. Camposinhos RS (2009) Study on the effect of cylindrical drilling of polyethylene sheaths. Study on the effect of "epdm" strips on continuous support – experimental analysis and computer modelling. Polytechnic of Porto – School of Engineering, Porto
11. Conroy K, Hoigard KR (2007) Stiffness considerations in dimension stone anchorage design. J ASTM Int 4(6):1–12. doi:JAI101062
12. Mises R (1945) On saint-Venant's principle. Bull Am Math Soc 51:555–562. doi:10.1090/S0002-9904-1945-08394-3
13. Pilkey WD, Pilkey DF, Peterson RE, Books24×7 Inc. (2008) Peterson's stress concentration factors, 3rd edn. John Wiley & Sons, Inc., Hoboken, New Jersey
14. Shah SP, Swartz SE, Ouyang C (1995) Fracture mechanics of concrete. Wiley, West Conshohocken
15. Sauoma V, Natekar D, Hansen (2003) Cohesive stresses and size effects in elasto-plastic and quasi-brittle materials. Int J Fract 119:287–298
16. Albrecht P, Yamada K (1977) Rapid calculation of stress intensity factors. J Struct Div ASCE 103(S.1):377–389
17. Rooke DP (1976) Compendium of stress intensity factors. H.M.S.O, London
18. ASTM – American Society for Testing Materials (2010) Standard guide for selection, design, and installation of dimension stone attachment systems, vol ASTM C1242 - 10. American Standard of Testing Materials, West Conshohocken

# Chapter 9
# Stainless Steel and Aluminium Anchors

**Abstract** Stainless steel and aluminium alloys are dealt exclusively as the main materials to use in anchors and supporting systems for stone cladding.

Special attention is given to bimetallic corrosion when dissimilar materials are used in aggressive environments.

Physical and mechanical properties of both metal alloys are provided in order to permit the designer to verify strength and deformation of the supporting elements and connections. Formula is presented in a unified way given the similarity of their ductile behaviour.

Design of members subjected to combined forces, such as bending and shear or shear and tension are illustrated making use of the dowel and pin, and the kerf anchor systems.

## 9.1 Introduction

The design of any anchor systems revolves about two main issues: the ability to transfer the loads to the building's façade structure without endangering the stone's cladding function and sustain enough strength to preserve the risk of failure at an appropriate level.

The theories for structural analysis, beam bending and structural mechanics are not considered in this chapter and are out of the scope of the book. Thus the reader should be familiar with the principles of structural engineering and mechanics.

An open rainscreen system is nowadays, without any doubt, the best solution for any façade technology system. Thus, the presence of water either in form of vapour or as a liquid is supposed to be present and in contact with the metallic supporting systems of any cladding. This is the main reason why designers and practitioners evidence the advantages of the employ of stainless steel or aluminium in both economic and efficiency terms.

R. de Sousa Camposinhos, *Stone Cladding Engineering*,
DOI 10.1007/978-94-007-6848-2_9, 

Saying that it must be assured that, in the accepted life cycle, deterioration of the metallic parts will not occur as long as strength and deformations limits states are granted.

Hence, in the following sections, some data and formulae are presented to allow the safety verification for the resistance and serviceability for the more used shapes and geometry either in stainless steel or aluminium alloys cladding supporting systems. Furthermore, the issues related with bimetallic corrosion are discussed.

## 9.2 Corrosion and Metallic Alloys

The Empire State Building's façade did not require repairs for almost 60 years of service, which represented an outstanding service life of its exterior wall system: a thin limestone facing that was anchored into a brick back-up masonry wall which in turn was tied into structural steel columns.

When repairs were being done, it was discovered that the iron anchors, which tied the brick back-up to the column, experienced significant corrosion due to oxidation. As the brick masonry had been built directly against the steel column, the expansive forces that were developed by corrosion displaced and cracked the masonry back-up, which resulted in cracking on the limestone façade.

Cracking allowed further water penetration into the wall system accelerating steel corrosion and masonry displacement. Repairs were attempted in the 1950s to contain movement in the limestone by installing steel straps around the outside corners, but this did little to slow down or stop the deterioration process [1].

Early systems rarely consider the effects of water penetrating the cladding system. Galvanized steel may have, in some circumstances, been used for connection components including shelf angles, lateral straps, and bolts. As the system aged, galvanized and unprotected steel would have eventually corroded and resulted in the failure of components or of the entire system.

Within the past 30 years, stainless steel has been recommended for all anchorage components that are in contact with the stone, yet, even components with high corrosion resistance may corrode if two different metals are in contact due to galvanic corrosion in which the rate of corrosion of the less noble metal increases.

Particular attention to galvanic corrosion is necessary in environments with airborne chlorides such as ocean properties. Metals in direct contact with stone must be corrosion resistant ones, yet in harsh environments some stone types may exhibit relevant degree of corrosion. Biotitic granite and other iron containing stones [2] are an example.

Attention should be given to bimetallic corrosion which can occur when two dissimilar metals are in contact and bridged by an electrically conductive liquid [3].

The more electrically conductive the liquid is, the greater is the danger of corrosion. Seawater or salt laden moist air is more of a risk than contact with rain water or town water. In fact, if the metals are dry, bimetallic (galvanic) corrosion cannot occur.

The combination of aluminium and stainless steel may be considered to have some, yet low, bi-metallic corrosion risk, although aluminium is anodic to stainless steel. Large relative surfaces of aluminium to stainless steel can be considered safe, for example where stainless steel fasteners and bolts are used to attach aluminium profiles.

In very severe environment, such as coastal and marine areas or even chemical industrial environment, stainless steel bolts and nuts with sound insulating washers can be used, as long as isolating washer are interposed so that breaking the corrosion is attained by isolating the two "dissimilar" metals in such conditions.

These aspects need to be taken into account from the very start of the anchors design process but as a practice, all dissimilar metals should not be in contact [4–6].

## 9.3 Aluminium Versus Steel

The behaviour of materials can be roughly classified into two categories: brittle and ductile. Steel and aluminium usually fall in the class of ductile materials. Ductile materials exhibit large strains and yielding before they fail, so that similarities between aluminium and steel may be pointed out such as having the same structural applications with very similar design approaches. Furthermore, the design rules purposed in worldwide codes are comparable.

However, there are important differences in physical as well as mechanical properties which have to be accounted for in the design process.

A comparison between some of the most important physical and mechanical properties of aluminium and steel is presented in Table 9.1.

## 9.4 Stainless Steel

Stainless steel differs from carbon steel by the amount of chromium present. Unprotected carbon steel rusts readily when exposed to air and moisture and accelerates corrosion by forming more iron oxide, and due to the dissimilar size of the iron and iron oxide molecules tend to flake and fall away. Stainless steels contain sufficient chromium to form a passive film of chromium oxide, which

**Table 9.1** Physical and mechanical properties of Aluminium and Steel

| | | Aluminium | Steel |
|---|---|---|---|
| Volumic mass | $kg/m^3$ | 2,700 | 7,800 |
| Young modulus | MPa | 70,000 | 210,000 |
| Shear modulus | MPa | 27,000 | 81,000 |
| Poisson ratio | – | 0.33 | 0.3 |
| Coefficient of linear thermal expansion | $K^{-1}$ | $23 \times 10^{-6}$ | $12 \times 10^{-6}$ |

**Table 9.2** Typical mechanical properties of stainless steel types

| Grade | $f_u$ – Ultimate tensile strength (MPa) | $f_{0.2}$ – 0.2 % proof strength (MPa) | Elongation % in 2″ | Hardness rockwell |
|---|---|---|---|---|
| 304 | 621 | 290 | 55 | B82 |
| 304 L | 586 | 241 | 55 | B80 |
| Type 316 | 579 | 290 | 50 | B79 |
| Type 316 L | 558 | 290 | 50 | B79 |

prevents further surface corrosion and blocks corrosion from spreading into the metal's structure, and due to the similar size of the steel and oxide molecules they bond very strongly and remain attached to the surface [7].

The most common grade of stainless steel for general usage is commonly referred to as "18–8", from its Chrome (nominal 18 %) and Nickel (nominal 8 %) content. These are austenitic non-magnetic resistant to most acids, and offer good corrosion resistance. The two types most generally used are the type 304 and type 316.

Type 304 has as principal elements, 18–20 % Chrome, 8–12 % Nickel with 0.08 % of maximum Carbon.

Type 316 has as principal elements, 16–18 % Chrome, 10–14 % Nickel together with 0.08 % maximum Carbon and 2 % maximum Molybdenum. This grade has greater corrosion resistance when exposed to harsh environments as compared to the general range of 18–8. This is due to the addition of Molybdenum and the higher percentage of Nickel.

Types 304L or 316L are an extra low-carbon variation of their originals with a 0.03 % maximum carbon content that eliminates carbide precipitation due to welding. As a result, these alloys can be used in the "as-welded" condition, even in severe corrosive conditions. It often eliminates the necessity of annealing weldments except for applications specifying stress relief. It has slightly lower mechanical properties than Type 304. Table 9.2 summarizes the principal mechanical characteristic values of these stainless steel grades.

Bolts, nuts, screws, cap screws, sheet metal screws, and other items are cold-headed or hot-forged and are available either in metric or imperial sizes.

### *9.4.1 Stainless Steel Bolts and Nuts*

The most common shape used for the bolt head or for the nuts is hexagonal which gives favourable angles for a tool to approach given that more (and smaller) corners would be susceptible to being rounded off.

The strength classes of stainless steel bolts are categorized according to the stainless-steel material types used. Their properties are categorized by strength classes, but with a different numbering system.

Table 9.3 shows a part of ISO 3506–1 that rules on mechanical properties of stainless steel bolts.

**Table 9.3** Mechanical properties of stainless steel bolts

| Strength class | $f_u$ – Ultimate tensile strength (MPa) | $f_{0.2}$ – 0.2 % proof strength (MPa) | $d_b$ bolt diameter (mm) |
|---|---|---|---|
| A2–50 | 500 | 210 | $d_b \leq 39$ |
| A4–70 | 700 | 450 | $d_b \leq 39$ |
| A4–80 | 800 | 600 | $d_b \leq 39$ |

From ISO3506-1 [8]

A2-50 grade is a general purpose stainless steel also known as grade 304 or 18/8. A4-70 grade stainless steel corresponds to the grade 316 and is, as referred, a higher corrosion resistant, marine grade of stainless steel.

The A4-80 grade stainless steel has identical corrosion resistance to A4-70 but with a higher tensile strength comparable to that of 8.8 high tensile steel.

Nuts are graded with strength ratings attuned with their respective bolts in order to be able to support the bolt proof strength load.

## 9.5 Aluminium Alloys

Pure aluminium is a weak material yet it can be strengthened by alloying and subsequent treatment. Aluminium alloys are grouped in eight series, from 1xxx to 8xxx and an addition difference is made depending on the heat or non-heat treatment [9].

The aluminium alloys mostly used in extruded shapes for architecture, particularly window frames, door frames, open shaped profiles, etc., belong to the 6xxx series. Containing magnesium and silicon, these alloys have very good extrudability and resistance to corrosion together with a high strength. Tensile strength has a value around 300 MPa with a proof stress of 250 MPa. This group includes the 6082 and 6063 series which are widely employed in building structures. Table 9.4 presents their mechanical properties.

It is typically produced with very smooth surfaces fit for anodizing, has generally good mechanical properties and is heat-treatable and weldable, although strength near the weld can be lost up to 30 % without subsequent heat treating.

The mechanical properties of these alloys can be noticeably changed by heat treatment with or without additional strain hardening or the so-called tempering. Temper T4 stands for solution heat-treated, cold-worked and naturally aged, T5 for cooled from an elevated temperature shaping process and artificially aged, and T6 for a solution heat-treated and artificially aged.

### *9.5.1 Aluminium Bolts and Nuts*

Despite the choices available for fastener materials, it is recommended that they should be manufactured from aluminium alloy to avoid thermal expansion problems and also the contact with more noble materials.

**Table 9.4** Mechanical properties of wrought aluminium alloys

| Alloy designation | Temper | $t$ –Thickness or wall thickness (mm) | $f_u$ – Ultimate tensile strength (MPa) | $f_{0.2}$ – 0.2 % proof strength (MPa) | Minimum elongation (%) |
|---|---|---|---|---|---|
| 6082 | T4 | $t \leq 25$ | 205 | 110 | 14 |
| 6082 | T5 | $t \leq 5$ | 270 | 230 | 8 |
| 6082 | T6 | $5 < t \leq 25$ | 310 | 260 | 10 |
| 6063 | T5 | $t \leq 3$ | 215 | 170 | 8 |
| 6063 | T5 | $3 < t \leq 25$ | 160 | 110 | 7 |
| 6063 | T6 | $t \leq 10$ | 215 | 170 | 8 |
| 6063 | T6 | $t \leq 20$ | 220 | 190 | 10 |
| 7020 | T4 | $t \leq 12.5$ | 265 | 210 | 12 |
| 7020 | T6 | $t \leq 12.5$ | 315 | 280 | 10 |
| 7020 | T4 | $t \leq 40$ | 320 | 290 | 8 |

Extracted from Eurocode 9 [10]

**Table 9.5** Mechanical properties of aluminium alloys bolts

| Alloy designation | Temper | $f_u$ – Ultimate tensile strength (MPa) | $f_{0.2}$ – 0.2 % proof strength (MPa) | $d_b$ Bolt nominal diameter (mm) |
|---|---|---|---|---|
| 2024 | T4 | 427 | 275 | $3 < d_b \leq 25$ |
| 5056 | H24 | 330 | 250 | $3 < d_b \leq 25$ |
| 6061 | T6 | 289 | 241 | $3 < d_b \leq 25$ |
| 6082 | T6 | 340 | 310 | $3 < d_b \leq 25$ |
| 6262 | T9 | 359 | 330 | $3 < d_b \leq 50$ |
| 6262 | T9 | 359 | 317 | $50 < d_b \leq 75$ |
| 7075 | T7 | 469 | 386 | $10 < d_b \leq 25$ |

Extracted from Aluminum Fastener Supply, Inc. Catalogue [11]

As for stainless steel, aluminium bolts and nuts are available in all common metric and imperial sizes, although aluminium bolts and nuts have a slight shortcoming in terms of strength when compared to stainless steel bolts; however compensating advantages include resistance to corrosion, lower than a stainless steel cost and naturally the compatibility to joint aluminium alloy elements. Table 9.5 summarizes for the most used aluminium alloy bolts the ultimate and proof tensile strength.

## 9.6 Anchors Design

Prior to designing a member, an analysis has to be carried out to determine its loading and properties. Bearing in mind that modern analysis software using three-dimensional models is frequently used to find interactions between the members, a simple, clear and effective solution may, in most cases, be achieved with prompt "hand calculations".

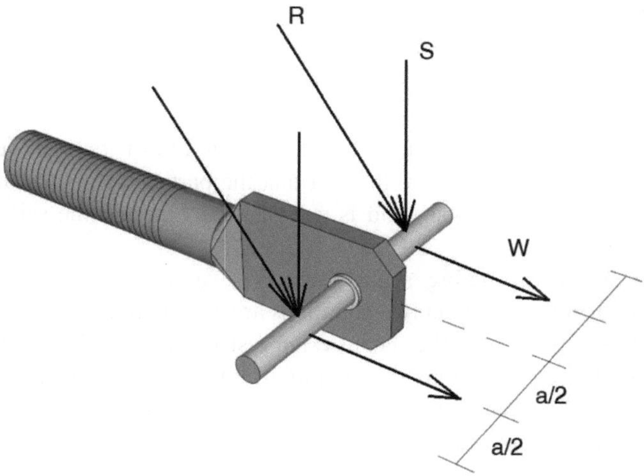


**Fig. 9.1** Schematic forces acting on a dowel pin; (*W*) wind action; (*S*) self-weight of the dimension stone; (*R*) resultant

Special consideration to aluminium material of almost all alloys should be taken in presence of welded members. Most of the aluminium alloys members sections are weakened in the heat affected zones adjacent to welds [12], which is not the case of stainless steel welded connections [13]. This aspect is not considered in the following sections where the design of members subjected to combined forces is approached.

### *9.6.1 Bending and Shear*

It's uncommon for a designer to find pure shear or pure bending even when simple connections are being analysed. We can take for example the schematic representation of Fig. 9.1 where the forces transmitted from the stone to a loose pin are illustrated. In this view it's considered that the pin is horizontal thus receiving the self-weight of the cladding.

It should be noted that due to the presence of a sleeve (not represented) the resultant of the applied forces may be reasonably assumed to be located at a distance equal to half of the pin length, (a/2), on each side of supporting bolt or shaft.

When the pin is in vertical bearing position only the wind has to be taken into account as the stone cladding weight is directly transferred to the bolt or shaft.

To verify the capacity of the pin it's necessary to observe that the design value of the resultant $R$ and of the inherent maximum bending moment value located at the cross section at the bolt or shaft insertion point.

Generally one has:

$$R = \sqrt{W^2 + S^2} \tag{9.1}$$

where $W$ and $S$ are design value of wind reaction and of the self-weight of the stone. Attention should be paid to the fact that the partial factors for these actions have different values since the wind is a variable action and the self-weight is a permanent action.

Denoting $R$ by the acting shear force, $V_{Sd}$, and $M_{Sd} = V_{Sd} \cdot a/2$, the corresponding bending moment two situations are considered:

If $V_{Sd}$ is less than 50 % of the shear design resistance, $V_{Rd}$, then no reduction in the design resistance moment, $M_{Rd}$, is required and independent safety verifications are permitted for the shear and bending moment resistance.

For the shear resistance the condition is:

$$V_{Rd} \geq V_{Sd} \tag{9.2}$$

with

$$V_{Rd} = \frac{A_v \cdot f_{0.2}}{\sqrt{3} \times \gamma_{M1}} \tag{9.3}$$

where:

$A_v$ is the shear area, $f_{0.2}$ is the characteristic value of 0.2 % proof strength of the stainless steel or the aluminium alloy and $\gamma_{M1}$ the partial safety facto for resistance taken with a value of 1.1.

The shear area is based on the shape of the cross section. For closed hollow or solid sections it corresponds to the cross section area.

For the bending resistance the condition is:

$$M_{Rd} \geq M_{Sd} \tag{9.4}$$

The value of the bending moment resistance should be calculated considering the net cross-sections at holes, if present, in order to compensate for the weakening effects, as follows:

$$M_{Rd} = \frac{W_{\text{net}} \cdot f_u}{\gamma_{M2}} \tag{9.5}$$

In general, for any cross-section, a plastic hinge can be formed and a plastic moment can be developed so that the bending moment resistance is given by the following expression:

$$M_{Rd} = \frac{W_{\text{pl}} \cdot f_{0.2}}{\gamma_{M1}} \tag{9.6}$$

In expression (9.5) and expression (9.6) $f_u$ is the ultimate tensile strength of the metal alloy, $f_{o.2}$ is the 0.2 % proof strength of the metal alloy, $W_{\text{net}}$ the elastic modulus of the net section allowing for holes, $W_{\text{pl}}$ is the plastic modulus of gross section, $\gamma_{M1} = 1.1$ the partial factor for general yielding and $\gamma_{M2} = 1.25$. It is important to realize that for aluminium alloy elements in the heat affected zones a reduction of the resistance shall be taken into account in the design formulae. Values of softening factor are given, for example in Table 5.2 of the Eurocode 9 [10].

When $V_{Sd}$ is higher than 50 % of the shear design resistance, $V_{Rd}$, the design resistance moment is reduced to:

$$M_{Rd.\text{red}} = \frac{f_{0v} \cdot W_{\text{el}}}{\gamma_{M1}} \tag{9.7}$$

with

$$f_{0v} = f_{0.2}\left[1 - \left(2\frac{V_{Sd}}{V_{Rd}} - 1\right)^2\right] \tag{9.8}$$

## 9.6.2 *Shear and Tension*

Mechanical connections may be achieved by bolting, screwing, riveting and pining and are frequently used as connection methods when joining metal members.

Bolting is the preferred method to mount the metal skeleton or framework for stone cladding support.

Design shear resistance of bolts per shear plane is given by:

$$F_{V,Rd} = \frac{0.5 \times f_u \cdot A}{\gamma_{M2}} \tag{9.9}$$

On the other hand, the tensile resistance is obtained as follows:

$$F_{T,Rd} = \frac{0.9 \times f_u \cdot A}{\gamma_{M2}} \tag{9.10}$$

In the above equations the same notation is used as for the previous ones. Attention should be paid to the value of the stress area of the cross section of the bolts, $A$, when the shear resistance is being determined in Eq. (9.9):

$A = A_s$ is the stress area of the cross section if the shear plane passes through the threaded portion of the bolt;

$A = A$ is the nominal area of the cross section if the shear plane passes through the unthreaded portion of the bolt;

The values of the stress area are given in Table 9.6.

**Table 9.6** Values of nominal area and stress area for common bolt diameters

| Nominal diameter $d_b$ (mm) | Nominal area A ($mm^2$) | Stress area $A_s$ ($mm^2$) |
|---|---|---|
| 8 | 50.3 | 36.6 |
| 10 | 78.5 | 58 |
| 12 | 113 | 84.3 |
| 14 | 154 | 115 |
| 16 | 201 | 157 |
| 18 | 254 | 192 |
| 20 | 314 | 245 |
| 22 | 380 | 303 |
| 24 | 452 | 353 |
| 27 | 573 | 459 |
| 30 | 707 | 561 |

However, bolts are mainly subject to combined shear and tension and their resistance must satisfy the following condition:

$$\frac{F_{V,Sd}}{F_{V,Rd}} + \frac{F_{T,Sd}}{1.4 \times F_{T,Rd}} \leq 1.0 \tag{9.11}$$

An example for this situation is illustrated in Fig. 9.2, where a split expansion sleeve over a threaded stud bolt body is used for anchorage into concrete of a cladding supporting rail.

### *9.6.3 Bending Tension and Shear*

Most often, the cross section of a member is subjected to several loadings simultaneously. As long as the relationship between stress and the loads is linear and the geometry of the member would not undergo significant change when the loads are applied, the principle of superposition can be applied.

Let us consider again the example of Fig. 9.2 and take as reference the cross section at a distance, $e_1$, from the centre line of the applied force S due to the self-weight of the stone per unit length of the rail. For the unit length of the cross-section and given, $t_p$, the thickness of the profile the combine loading is established given the axial force due to the wind load, the bending moment and shear reactions due to the stone cladding self-weight.

These loads are resolved into components parallel and transverse to the longitudinal axis of the rail and the corresponding stresses are calculated:

$$\begin{aligned} \sigma_{Sd} &= 6 \times \frac{M_{Sd}}{t_p^2} + \frac{N_{Sd}}{t_p}; \\ \tau_{Sd} &= \frac{V_{Sd}}{t_p} \end{aligned} \tag{9.12}$$

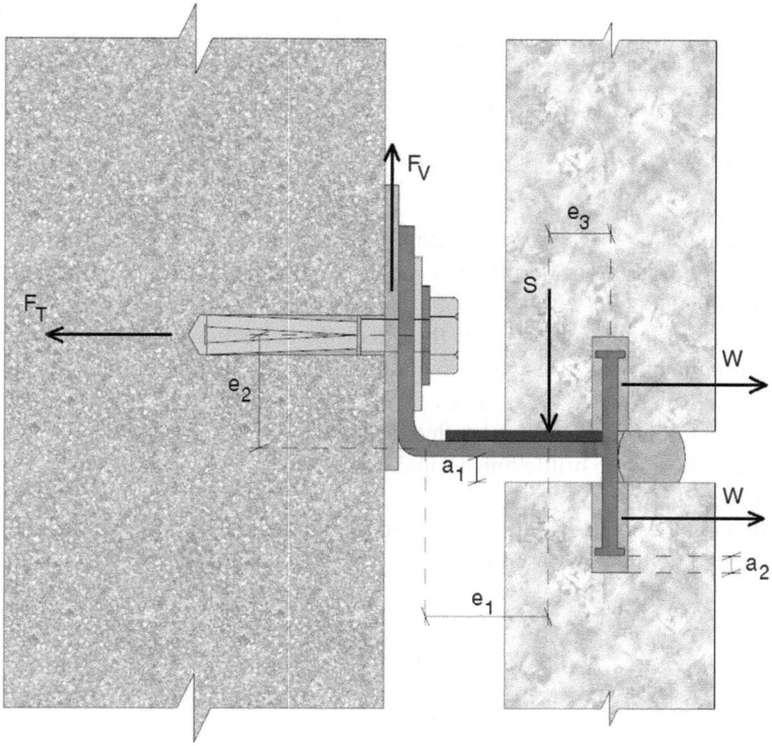


**Fig. 9.2** Schematic forces acting on an anchorage to concrete; (*W*) wind action; (*S*) self-weight of the dimension stone; ($F_T$) tensile reacting force; ($F_V$) shear reacting force; ($e_1$) cladding self-weight eccentricity; ($e_2$) bolt eccentricity;($a_1$) and ($a_2$) cleat deformation clearances

with

$$M_{Sd} = \gamma_g \cdot S \cdot e_1; \quad V_{Sd} = \gamma_g \cdot S; \quad N_{Sd} = 2 \times \gamma_f \cdot W \tag{9.13}$$

Application of the von Mises criterion to these stress components gives that the equivalent stress in the cross-section satisfies the relation:

$$\sqrt{\sigma_{Sd}^2 + 3 \times (\tau_{Sd}^2)} \leq \frac{f_u}{\gamma_{M1}} \tag{9.14}$$

Substitution of Eq. (9.12) in expressions (9.14) defines an equation allowing for the safety verification of the anchor resistance in that cross-section.

As a conservative approximation a linear summation of the utilisation ratios for each stress resultant may be used. For cross sections subjected to the combination of $N_{Sd}$ and $M_{Sd}$ the following criteria may be applied:

$$\frac{N_{Sd}}{N_{Rd}} + \frac{M_{Sd}}{M_{Rd}} \leq 1 \tag{9.15}$$

where $N_{Rd}$ and $M_{Rd}$ are the design values of the resistance including any reduction that may be caused by shear effects as stated in Eq. (9.7).

### 9.6.4 Deformation Limit State

Displacements in horizontal supporting elements, when loaded vertically or laterally, are supposed to occur, although within specified limits to avoid malfunctions and cladding damage.

With lesser modulus of elasticity than steel, the deflection of aluminium elements can be dominant in the design. Albeit suitable limits can be agreed between the designer and the owner, *e.g.*, values for $a_1$ or $a_2$ in Fig. 9.2, as a general rule in curtain wall mullions the vertical deflection should be limited to L/250 or 15 mm and in cantilevers carrying brittle finish a maximum deflection of L/360 is recommended in Eurocode 9 [10].

Following the same example of the kerf cladding from Fig. 9.2 the deflections to be compared with the imposed values may be checked as follows:

$$\begin{aligned} a_1 &\geq \frac{S \cdot e_1^3}{3 \times E \cdot I} = 4 \times \frac{S \cdot e_1^3}{E \cdot t_p^3} \\ a_2 &\geq \frac{S \cdot e_1^2 \cdot (2 \times e_1 + 3 \times e_3)}{6 \times E \cdot I} = 2 \times \frac{S \cdot e_1^2 \cdot (2 \times e_1 + 3 \times e_3)}{E \cdot t_p^3} \end{aligned} \tag{9.16}$$

with

$S$ the characteristic value per unit of length of the cladding stone, $E$ the modulus of elasticity of the metal alloy and $t_p$ the thickness of the rail profile.

## References


1. Nacheman RJ (2006) The empire state building – façade evaluation and repair of an engineering landmark. Structure. C3 Ink, Wisconsin
2. Camposinhos RS, Pires V, Simão J, Garcia ML (2009) Patologia em Pedra Natural. Manchas em revestimento de Granito Amarelo Luzelos. Paper presented at the 3.º Encontro Sobre Patologia Reabilitação de Edifícios, Porto
3. Materials A-ASfTa (2010) Standard guide for selection, design, and installation of dimension stone attachment systems, vol ASTM C1242 – 10. American Standard of Testing Materials, West Conshohocken
4. Borgal C (2001) Thin-stone systems: conflicts between reality and expectations. APT Assoc Preserv Tech Int 32(1):19–25
5. Scheffler MJ (2001) Thin-stone veneer building facades: evolution and preservation. APT Assoc Preserv Tech Int 32(1):27–34
6. Gerns EA (2002) Design issues for thin-stone cladding systems. Paper presented at the symposium of building envelope technology, Coral Springs, FL, 14–15, 2002

7. Qiu J (1995) Stainless steels and alloys: why they resist corrosion and how they fail. Corrosion special topical papers. Nanyang Technological University, Singapore
8. ISO (2009) ISO 3506–1 Mechanical properties of corrosion-resistant stainless steel fasteners. Part 1: Bolts, screws and studs
9. The Aluminum Association (2009) International alloy designations and chemical composition limits for wrought aluminum and wrought aluminum alloys. The Aluminum Association, Arlington
10. Staff BSI, Institution BS (2000) Eurocode 9. Design of aluminium structures. General rules. General rules and rules for buildings. B S I Standards
11. Aluminum Fastener Supply Co. Inc. (2012) Product catalog. Aluminum Fastener Supply Co., Inc., Naples
12. Dwight JB (1999) Aluminium design and construction. E & FN Spon
13. The Steel Construction Institute (2003) Design manual for structural stainless steel – commentary. Euro Inox and The Steel Construction Institute Ascot, Ascot

# Index



R. de Sousa Camposinhos, *Stone Cladding Engineering*,
DOI 10.1007/978-94-007-6848-2, © Springer Science+Business Media Dordrecht 2014

MIX
Papier aus verantwortungsvollen Quellen
Paper from responsible sources
FSC® C105338
FSC
www.fsc.org

If you have any concerns about our products,
you can contact us on
**ProductSafety@springernature.com**

In case Publisher is established outside the EU,
the EU authorized representative is:
**Springer Nature Customer Service Center GmbH**
**Europaplatz 3, 69115 Heidelberg, Germany**

Printed by Libri Plureos GmbH
in Hamburg, Germany